Geometry: A Very Short Introduction

VERY SHORT INTRODUCTIONS are for anyone wanting a stimulating and accessible way into a new subject. They are written by experts, and have been translated into more than 45 different languages.

The Series began in 1995, and now covers a wide variety of topics in every discipline. The VSI library currently contains over 700 volumes—a Very Short Introduction to everything from Psychology and Philosophy of Science to American History and Relativity—and continues to grow in every subject area.

Very Short Introductions available now:

ABOLITIONISM Richard S. Newman
THE ABRAHAMIC RELIGIONS Charles L. Cohen
ACCOUNTING Christopher Nobes
ADOLESCENCE Peter K. Smith
ADVERTISING Winston Fletcher
AERIAL WARFARE Frank Ledwidge
AESTHETICS Bence Nanay
AFRICAN AMERICAN RELIGION Eddie S. Glaude Jr
AFRICAN HISTORY John Parker and Richard Rathbone
AFRICAN POLITICS Ian Taylor
AFRICAN RELIGIONS Jacob K. Olupona
AGEING Nancy A. Pachana
AGNOSTICISM Robin Le Poidevin
AGRICULTURE Paul Brassley and Richard Soffe
ALEXANDER THE GREAT Hugh Bowden
ALGEBRA Peter M. Higgins
AMERICAN BUSINESS HISTORY Walter A. Friedman
AMERICAN CULTURAL HISTORY Eric Avila
AMERICAN FOREIGN RELATIONS Andrew Preston
AMERICAN HISTORY Paul S. Boyer
AMERICAN IMMIGRATION David A. Gerber
AMERICAN INTELLECTUAL HISTORY Jennifer Ratner-Rosenhagen
AMERICAN LEGAL HISTORY G. Edward White
AMERICAN MILITARY HISTORY Joseph T. Glatthaar
AMERICAN NAVAL HISTORY Craig L. Symonds
AMERICAN POETRY David Caplan
AMERICAN POLITICAL HISTORY Donald Critchlow
AMERICAN POLITICAL PARTIES AND ELECTIONS L. Sandy Maisel
AMERICAN POLITICS Richard M. Valelly
THE AMERICAN PRESIDENCY Charles O. Jones
THE AMERICAN REVOLUTION Robert J. Allison
AMERICAN SLAVERY Heather Andrea Williams
THE AMERICAN SOUTH Charles Reagan Wilson
THE AMERICAN WEST Stephen Aron
AMERICAN WOMEN'S HISTORY Susan Ware
AMPHIBIANS T. S. Kemp
ANAESTHESIA Aidan O'Donnell
ANALYTIC PHILOSOPHY Michael Beaney
ANARCHISM Colin Ward
ANCIENT ASSYRIA Karen Radner
ANCIENT EGYPT Ian Shaw
ANCIENT EGYPTIAN ART AND ARCHITECTURE Christina Riggs
ANCIENT GREECE Paul Cartledge

THE ANCIENT NEAR EAST Amanda H. Podany
ANCIENT PHILOSOPHY Julia Annas
ANCIENT WARFARE Harry Sidebottom
ANGELS David Albert Jones
ANGLICANISM Mark Chapman
THE ANGLO-SAXON AGE John Blair
ANIMAL BEHAVIOUR Tristram D. Wyatt
THE ANIMAL KINGDOM Peter Holland
ANIMAL RIGHTS David DeGrazia
THE ANTARCTIC Klaus Dodds
ANTHROPOCENE Erle C. Ellis
ANTISEMITISM Steven Beller
ANXIETY Daniel Freeman and Jason Freeman
THE APOCRYPHAL GOSPELS Paul Foster
APPLIED MATHEMATICS Alain Goriely
THOMAS AQUINAS Fergus Kerr
ARBITRATION Thomas Schultz and Thomas Grant
ARCHAEOLOGY Paul Bahn
ARCHITECTURE Andrew Ballantyne
THE ARCTIC Klaus Dodds and Jamie Woodward
ARISTOCRACY William Doyle
ARISTOTLE Jonathan Barnes
ART HISTORY Dana Arnold
ART THEORY Cynthia Freeland
ARTIFICIAL INTELLIGENCE Margaret A. Boden
ASIAN AMERICAN HISTORY Madeline Y. Hsu
ASTROBIOLOGY David C. Catling
ASTROPHYSICS James Binney
ATHEISM Julian Baggini
THE ATMOSPHERE Paul I. Palmer
AUGUSTINE Henry Chadwick
JANE AUSTEN Tom Keymer
AUSTRALIA Kenneth Morgan
AUTISM Uta Frith
AUTOBIOGRAPHY Laura Marcus
THE AVANT GARDE David Cottington
THE AZTECS Davíd Carrasco
BABYLONIA Trevor Bryce
BACTERIA Sebastian G. B. Amyes
BANKING John Goddard and John O. S. Wilson
BARTHES Jonathan Culler
THE BEATS David Sterritt
BEAUTY Roger Scruton
BEHAVIOURAL ECONOMICS Michelle Baddeley
BESTSELLERS John Sutherland
THE BIBLE John Riches
BIBLICAL ARCHAEOLOGY Eric H. Cline
BIG DATA Dawn E. Holmes
BIOCHEMISTRY Mark Lorch
BIOGEOGRAPHY Mark V. Lomolino
BIOGRAPHY Hermione Lee
BIOMETRICS Michael Fairhurst
BLACK HOLES Katherine Blundell
BLASPHEMY Yvonne Sherwood
BLOOD Chris Cooper
THE BLUES Elijah Wald
THE BODY Chris Shilling
THE BOOK OF COMMON PRAYER Brian Cummings
THE BOOK OF MORMON Terryl Givens
BORDERS Alexander C. Diener and Joshua Hagen
THE BRAIN Michael O'Shea
BRANDING Robert Jones
THE BRICS Andrew F. Cooper
THE BRITISH CONSTITUTION Martin Loughlin
THE BRITISH EMPIRE Ashley Jackson
BRITISH POLITICS Tony Wright
BUDDHA Michael Carrithers
BUDDHISM Damien Keown
BUDDHIST ETHICS Damien Keown
BYZANTIUM Peter Sarris
CALVINISM Jon Balserak
ALBERT CAMUS Oliver Gloag
CANADA Donald Wright
CANCER Nicholas James
CAPITALISM James Fulcher
CATHOLICISM Gerald O'Collins
CAUSATION Stephen Mumford and Rani Lill Anjum
THE CELL Terence Allen and Graham Cowling
THE CELTS Barry Cunliffe

CHAOS Leonard Smith
GEOFFREY CHAUCER David Wallace
CHEMISTRY Peter Atkins
CHILD PSYCHOLOGY Usha Goswami
CHILDREN'S LITERATURE
Kimberley Reynolds
CHINESE LITERATURE Sabina Knight
CHOICE THEORY Michael Allingham
CHRISTIAN ART Beth Williamson
CHRISTIAN ETHICS D. Stephen Long
CHRISTIANITY Linda Woodhead
CIRCADIAN RHYTHMS
Russell Foster and Leon Kreitzman
CITIZENSHIP Richard Bellamy
CITY PLANNING Carl Abbott
CIVIL ENGINEERING
David Muir Wood
CLASSICAL LITERATURE
William Allan
CLASSICAL MYTHOLOGY
Helen Morales
CLASSICS Mary Beard and
John Henderson
CLAUSEWITZ Michael Howard
CLIMATE Mark Maslin
CLIMATE CHANGE Mark Maslin
CLINICAL PSYCHOLOGY Susan
Llewelyn and Katie Aafjes-van Doorn
COGNITIVE NEUROSCIENCE
Richard Passingham
THE COLD WAR Robert J. McMahon
COLONIAL AMERICA Alan Taylor
COLONIAL LATIN AMERICAN
LITERATURE Rolena Adorno
COMBINATORICS Robin Wilson
COMEDY Matthew Bevis
COMMUNISM Leslie Holmes
COMPARATIVE LITERATURE
Ben Hutchinson
COMPETITION AND ANTITRUST
LAW Ariel Ezrachi
COMPLEXITY John H. Holland
THE COMPUTER Darrel Ince
COMPUTER SCIENCE
Subrata Dasgupta
CONCENTRATION CAMPS
Dan Stone
CONFUCIANISM Daniel K. Gardner
THE CONQUISTADORS
Matthew Restall and
Felipe Fernández-Armesto
CONSCIENCE Paul Strohm
CONSCIOUSNESS Susan Blackmore
CONTEMPORARY ART
Julian Stallabrass
CONTEMPORARY FICTION
Robert Eaglestone
CONTINENTAL PHILOSOPHY
Simon Critchley
COPERNICUS Owen Gingerich
CORAL REEFS Charles Sheppard
CORPORATE SOCIAL
RESPONSIBILITY Jeremy Moon
CORRUPTION Leslie Holmes
COSMOLOGY Peter Coles
COUNTRY MUSIC Richard Carlin
CREATIVITY Vlad Glăveanu
CRIME FICTION Richard Bradford
CRIMINAL JUSTICE
Julian V. Roberts
CRIMINOLOGY Tim Newburn
CRITICAL THEORY
Stephen Eric Bronner
THE CRUSADES
Christopher Tyerman
CRYPTOGRAPHY Fred Piper and
Sean Murphy
CRYSTALLOGRAPHY A. M. Glazer
THE CULTURAL REVOLUTION
Richard Curt Kraus
DADA AND SURREALISM
David Hopkins
DANTE Peter Hainsworth and
David Robey
DARWIN Jonathan Howard
THE DEAD SEA SCROLLS
Timothy H. Lim
DECADENCE David Weir
DECOLONIZATION Dane Kennedy
DEMENTIA Kathleen Taylor
DEMOCRACY Bernard Crick
DEMOGRAPHY Sarah Harper
DEPRESSION Jan Scott and
Mary Jane Tacchi
DERRIDA Simon Glendinning
DESCARTES Tom Sorell
DESERTS Nick Middleton
DESIGN John Heskett
DEVELOPMENT Ian Goldin
DEVELOPMENTAL BIOLOGY
Lewis Wolpert
THE DEVIL Darren Oldridge

DIASPORA Kevin Kenny
CHARLES DICKENS Jenny Hartley
DICTIONARIES Lynda Mugglestone
DINOSAURS David Norman
DIPLOMATIC HISTORY
Joseph M. Siracusa
DOCUMENTARY FILM
Patricia Aufderheide
DREAMING J. Allan Hobson
DRUGS Les Iversen
DRUIDS Barry Cunliffe
DYNASTY Jeroen Duindam
DYSLEXIA Margaret J. Snowling
EARLY MUSIC Thomas Forrest Kelly
THE EARTH Martin Redfern
EARTH SYSTEM SCIENCE Tim Lenton
ECOLOGY Jaboury Ghazoul
ECONOMICS Partha Dasgupta
EDUCATION Gary Thomas
EGYPTIAN MYTH Geraldine Pinch
EIGHTEENTHCENTURY BRITAIN
Paul Langford
THE ELEMENTS Philip Ball
EMOTION Dylan Evans
EMPIRE Stephen Howe
ENERGY SYSTEMS Nick Jenkins
ENGELS Terrell Carver
ENGINEERING David Blockley
THE ENGLISH LANGUAGE
Simon Horobin
ENGLISH LITERATURE Jonathan Bate
THE ENLIGHTENMENT
John Robertson
ENTREPRENEURSHIP Paul Westhead
and Mike Wright
ENVIRONMENTAL ECONOMICS
Stephen Smith
ENVIRONMENTAL ETHICS
Robin Attfield
ENVIRONMENTAL LAW
Elizabeth Fisher
ENVIRONMENTAL POLITICS
Andrew Dobson
ENZYMES Paul Engel
EPICUREANISM Catherine Wilson
EPIDEMIOLOGY Rodolfo Saracci
ETHICS Simon Blackburn
ETHNOMUSICOLOGY Timothy Rice
THE ETRUSCANS Christopher Smith
EUGENICS Philippa Levine
THE EUROPEAN UNION
Simon Usherwood and John Pinder
EUROPEAN UNION LAW
Anthony Arnull
EVOLUTION Brian and
Deborah Charlesworth
EXISTENTIALISM Thomas Flynn
EXPLORATION Stewart A. Weaver
EXTINCTION Paul B. Wignall
THE EYE Michael Land
FAIRY TALE Marina Warner
FAMILY LAW Jonathan Herring
MICHAEL FARADAY
Frank A. J. L. James
FASCISM Kevin Passmore
FASHION Rebecca Arnold
FEDERALISM Mark J. Rozell and
Clyde Wilcox
FEMINISM Margaret Walters
FILM Michael Wood
FILM MUSIC Kathryn Kalinak
FILM NOIR James Naremore
FIRE Andrew C. Scott
THE FIRST WORLD WAR
Michael Howard
FOLK MUSIC Mark Slobin
FOOD John Krebs
FORENSIC PSYCHOLOGY
David Canter
FORENSIC SCIENCE Jim Fraser
FORESTS Jaboury Ghazoul
FOSSILS Keith Thomson
FOUCAULT Gary Gutting
THE FOUNDING FATHERS
R. B. Bernstein
FRACTALS Kenneth Falconer
FREE SPEECH Nigel Warburton
FREE WILL Thomas Pink
FREEMASONRY Andreas Önnerfors
FRENCH LITERATURE John D. Lyons
FRENCH PHILOSOPHY
Stephen Gaukroger and Knox Peden
THE FRENCH REVOLUTION
William Doyle
FREUD Anthony Storr
FUNDAMENTALISM Malise Ruthven
FUNGI Nicholas P. Money
THE FUTURE Jennifer M. Gidley
GALAXIES John Gribbin
GALILEO Stillman Drake

GAME THEORY Ken Binmore
GANDHI Bhikhu Parekh
GARDEN HISTORY Gordon Campbell
GENES Jonathan Slack
GENIUS Andrew Robinson
GENOMICS John Archibald
GEOGRAPHY John Matthews and David Herbert
GEOLOGY Jan Zalasiewicz
GEOMETRY Maciej Dunajski
GEOPHYSICS William Lowrie
GEOPOLITICS Klaus Dodds
GERMAN LITERATURE Nicholas Boyle
GERMAN PHILOSOPHY Andrew Bowie
THE GHETTO Bryan Cheyette
GLACIATION David J. A. Evans
GLOBAL CATASTROPHES Bill McGuire
GLOBAL ECONOMIC HISTORY Robert C. Allen
GLOBAL ISLAM Nile Green
GLOBALIZATION Manfred B. Steger
GOD John Bowker
GOETHE Ritchie Robertson
THE GOTHIC Nick Groom
GOVERNANCE Mark Bevir
GRAVITY Timothy Clifton
THE GREAT DEPRESSION AND THE NEW DEAL Eric Rauchway
HABEAS CORPUS Amanda Tyler
HABERMAS James Gordon Finlayson
THE HABSBURG EMPIRE Martyn Rady
HAPPINESS Daniel M. Haybron
THE HARLEM RENAISSANCE Cheryl A. Wall
THE HEBREW BIBLE AS LITERATURE Tod Linafelt
HEGEL Peter Singer
HEIDEGGER Michael Inwood
THE HELLENISTIC AGE Peter Thonemann
HEREDITY John Waller
HERMENEUTICS Jens Zimmermann
HERODOTUS Jennifer T. Roberts
HIEROGLYPHS Penelope Wilson
HINDUISM Kim Knott
HISTORY John H. Arnold
THE HISTORY OF ASTRONOMY Michael Hoskin
THE HISTORY OF CHEMISTRY William H. Brock
THE HISTORY OF CHILDHOOD James Marten
THE HISTORY OF CINEMA Geoffrey Nowell-Smith
THE HISTORY OF LIFE Michael Benton
THE HISTORY OF MATHEMATICS Jacqueline Stedall
THE HISTORY OF MEDICINE William Bynum
THE HISTORY OF PHYSICS J. L. Heilbron
THE HISTORY OF POLITICAL THOUGHT Richard Whatmore
THE HISTORY OF TIME Leofranc HolfordStrevens
HIV AND AIDS Alan Whiteside
HOBBES Richard Tuck
HOLLYWOOD Peter Decherney
THE HOLY ROMAN EMPIRE Joachim Whaley
HOME Michael Allen Fox
HOMER Barbara Graziosi
HORMONES Martin Luck
HORROR Darryl Jones
HUMAN ANATOMY Leslie Klenerman
HUMAN EVOLUTION Bernard Wood
HUMAN PHYSIOLOGY Jamie A. Davies
HUMAN RIGHTS Andrew Clapham
HUMANISM Stephen Law
HUME James A. Harris
HUMOUR Noël Carroll
THE ICE AGE Jamie Woodward
IDENTITY Florian Coulmas
IDEOLOGY Michael Freeden
THE IMMUNE SYSTEM Paul Klenerman
INDIAN CINEMA Ashish Rajadhyaksha
INDIAN PHILOSOPHY Sue Hamilton
THE INDUSTRIAL REVOLUTION Robert C. Allen
INFECTIOUS DISEASE Marta L. Wayne and Benjamin M. Bolker

INFINITY Ian Stewart
INFORMATION Luciano Floridi
INNOVATION Mark Dodgson and David Gann
INTELLECTUAL PROPERTY Siva Vaidhyanathan
INTELLIGENCE Ian J. Deary
INTERNATIONAL LAW Vaughan Lowe
INTERNATIONAL MIGRATION Khalid Koser
INTERNATIONAL RELATIONS Christian Reus-Smit
INTERNATIONAL SECURITY Christopher S. Browning
IRAN Ali M. Ansari
ISLAM Malise Ruthven
ISLAMIC HISTORY Adam Silverstein
ISLAMIC LAW Mashood A. Baderin
ISOTOPES Rob Ellam
ITALIAN LITERATURE Peter Hainsworth and David Robey
HENRY JAMES Susan L. Mizruchi
JESUS Richard Bauckham
JEWISH HISTORY David N. Myers
JEWISH LITERATURE Ilan Stavans
JOURNALISM Ian Hargreaves
JAMES JOYCE Colin MacCabe
JUDAISM Norman Solomon
JUNG Anthony Stevens
KABBALAH Joseph Dan
KAFKA Ritchie Robertson
KANT Roger Scruton
KEYNES Robert Skidelsky
KIERKEGAARD Patrick Gardiner
KNOWLEDGE Jennifer Nagel
THE KORAN Michael Cook
KOREA Michael J. Seth
LAKES Warwick F. Vincent
LANDSCAPE ARCHITECTURE Ian H. Thompson
LANDSCAPES AND GEOMORPHOLOGY Andrew Goudie and Heather Viles
LANGUAGES Stephen R. Anderson
LATE ANTIQUITY Gillian Clark
LAW Raymond Wacks
THE LAWS OF THERMODYNAMICS Peter Atkins
LEADERSHIP Keith Grint
LEARNING Mark Haselgrove
LEIBNIZ Maria Rosa Antognazza
C. S. LEWIS James Como
LIBERALISM Michael Freeden
LIGHT Ian Walmsley
LINCOLN Allen C. Guelzo
LINGUISTICS Peter Matthews
LITERARY THEORY Jonathan Culler
LOCKE John Dunn
LOGIC Graham Priest
LOVE Ronald de Sousa
MARTIN LUTHER Scott H. Hendrix
MACHIAVELLI Quentin Skinner
MADNESS Andrew Scull
MAGIC Owen Davies
MAGNA CARTA Nicholas Vincent
MAGNETISM Stephen Blundell
MALTHUS Donald Winch
MAMMALS T. S. Kemp
MANAGEMENT John Hendry
NELSON MANDELA Elleke Boehmer
MAO Delia Davin
MARINE BIOLOGY Philip V. Mladenov
MARKETING Kenneth Le Meunier-FitzHugh
THE MARQUIS DE SADE John Phillips
MARTYRDOM Jolyon Mitchell
MARX Peter Singer
MATERIALS Christopher Hall
MATHEMATICAL FINANCE Mark H. A. Davis
MATHEMATICS Timothy Gowers
MATTER Geoff Cottrell
THE MAYA Matthew Restall and Amara Solari
THE MEANING OF LIFE Terry Eagleton
MEASUREMENT David Hand
MEDICAL ETHICS Michael Dunn and Tony Hope
MEDICAL LAW Charles Foster
MEDIEVAL BRITAIN John Gillingham and Ralph A. Griffiths
MEDIEVAL LITERATURE Elaine Treharne
MEDIEVAL PHILOSOPHY John Marenbon
MEMORY Jonathan K. Foster
METAPHYSICS Stephen Mumford

METHODISM William J. Abraham
THE MEXICAN REVOLUTION Alan Knight
MICROBIOLOGY Nicholas P. Money
MICROECONOMICS Avinash Dixit
MICROSCOPY Terence Allen
THE MIDDLE AGES Miri Rubin
MILITARY JUSTICE Eugene R. Fidell
MILITARY STRATEGY Antulio J. Echevarria II
MINERALS David Vaughan
MIRACLES Yujin Nagasawa
MODERN ARCHITECTURE Adam Sharr
MODERN ART David Cottington
MODERN BRAZIL Anthony W. Pereira
MODERN CHINA Rana Mitter
MODERN DRAMA Kirsten E. Shepherd-Barr
MODERN FRANCE Vanessa R. Schwartz
MODERN INDIA Craig Jeffrey
MODERN IRELAND Senia Pašeta
MODERN ITALY Anna Cento Bull
MODERN JAPAN Christopher Goto-Jones
MODERN LATIN AMERICAN LITERATURE Roberto González Echevarría
MODERN WAR Richard English
MODERNISM Christopher Butler
MOLECULAR BIOLOGY Aysha Divan and Janice A. Royds
MOLECULES Philip Ball
MONASTICISM Stephen J. Davis
THE MONGOLS Morris Rossabi
MONTAIGNE William M. Hamlin
MOONS David A. Rothery
MORMONISM Richard Lyman Bushman
MOUNTAINS Martin F. Price
MUHAMMAD Jonathan A. C. Brown
MULTICULTURALISM Ali Rattansi
MULTILINGUALISM John C. Maher
MUSIC Nicholas Cook
MYTH Robert A. Segal
NAPOLEON David Bell
THE NAPOLEONIC WARS Mike Rapport
NATIONALISM Steven Grosby
NATIVE AMERICAN LITERATURE Sean Teuton
NAVIGATION Jim Bennett
NAZI GERMANY Jane Caplan
NEOLIBERALISM Manfred B. Steger and Ravi K. Roy
NETWORKS Guido Caldarelli and Michele Catanzaro
THE NEW TESTAMENT Luke Timothy Johnson
THE NEW TESTAMENT AS LITERATURE Kyle Keefer
NEWTON Robert Iliffe
NIELS BOHR J. L. Heilbron
NIETZSCHE Michael Tanner
NINETEENTHCENTURY BRITAIN Christopher Harvie and H. C. G. Matthew
THE NORMAN CONQUEST George Garnett
NORTH AMERICAN INDIANS Theda Perdue and Michael D. Green
NORTHERN IRELAND Marc Mulholland
NOTHING Frank Close
NUCLEAR PHYSICS Frank Close
NUCLEAR POWER Maxwell Irvine
NUCLEAR WEAPONS Joseph M. Siracusa
NUMBER THEORY Robin Wilson
NUMBERS Peter M. Higgins
NUTRITION David A. Bender
OBJECTIVITY Stephen Gaukroger
OCEANS Dorrik Stow
THE OLD TESTAMENT Michael D. Coogan
THE ORCHESTRA D. Kern Holoman
ORGANIC CHEMISTRY Graham Patrick
ORGANIZATIONS Mary Jo Hatch
ORGANIZED CRIME Georgios A. Antonopoulos and Georgios Papanicolaou
ORTHODOX CHRISTIANITY A. Edward Siecienski
OVID Llewelyn Morgan
PAGANISM Owen Davies
PAKISTAN Pippa Virdee
THE PALESTINIAN-ISRAELI CONFLICT Martin Bunton

PANDEMICS Christian W. McMillen
PARTICLE PHYSICS Frank Close
PAUL E. P. Sanders
PEACE Oliver P. Richmond
PENTECOSTALISM William K. Kay
PERCEPTION Brian Rogers
THE PERIODIC TABLE Eric R. Scerri
PHILOSOPHICAL METHOD
Timothy Williamson
PHILOSOPHY Edward Craig
PHILOSOPHY IN THE ISLAMIC WORLD Peter Adamson
PHILOSOPHY OF BIOLOGY
Samir Okasha
PHILOSOPHY OF LAW
Raymond Wacks
PHILOSOPHY OF MIND
Barbara Gail Montero
PHILOSOPHY OF PHYSICS
David Wallace
PHILOSOPHY OF SCIENCE
Samir Okasha
PHILOSOPHY OF RELIGION
Tim Bayne
PHOTOGRAPHY Steve Edwards
PHYSICAL CHEMISTRY Peter Atkins
PHYSICS Sidney Perkowitz
PILGRIMAGE Ian Reader
PLAGUE Paul Slack
PLANETARY SYSTEMS
Raymond T. Pierrehumbert
PLANETS David A. Rothery
PLANTS Timothy Walker
PLATE TECTONICS Peter Molnar
PLATO Julia Annas
POETRY Bernard O'Donoghue
POLITICAL PHILOSOPHY
David Miller
POLITICS Kenneth Minogue
POPULISM Cas Mudde and Cristóbal Rovira Kaltwasser
POSTCOLONIALISM Robert Young
POSTMODERNISM
Christopher Butler
POSTSTRUCTURALISM
Catherine Belsey
POVERTY Philip N. Jefferson
PREHISTORY Chris Gosden
PRESOCRATIC PHILOSOPHY
Catherine Osborne
PRIVACY Raymond Wacks
PROBABILITY John Haigh
PROGRESSIVISM Walter Nugent
PROHIBITION W. J. Rorabaugh
PROJECTS Andrew Davies
PROTESTANTISM Mark A. Noll
PSYCHIATRY Tom Burns
PSYCHOANALYSIS Daniel Pick
PSYCHOLOGY Gillian Butler and Freda McManus
PSYCHOLOGY OF MUSIC
Elizabeth Hellmuth Margulis
PSYCHOPATHY Essi Viding
PSYCHOTHERAPY Tom Burns and Eva Burns-Lundgren
PUBLIC ADMINISTRATION
Stella Z. Theodoulou and Ravi K. Roy
PUBLIC HEALTH Virginia Berridge
PURITANISM Francis J. Bremer
THE QUAKERS Pink Dandelion
QUANTUM THEORY
John Polkinghorne
RACISM Ali Rattansi
RADIOACTIVITY Claudio Tuniz
RASTAFARI Ennis B. Edmonds
READING Belinda Jack
THE REAGAN REVOLUTION
Gil Troy
REALITY Jan Westerhoff
RECONSTRUCTION Allen C. Guelzo
THE REFORMATION Peter Marshall
REFUGEES Gil Loescher
RELATIVITY Russell Stannard
RELIGION Thomas A. Tweed
RELIGION IN AMERICA
Timothy Beal
THE RENAISSANCE Jerry Brotton
RENAISSANCE ART
Geraldine A. Johnson
RENEWABLE ENERGY Nick Jelley
REPTILES T. S. Kemp
REVOLUTIONS Jack A. Goldstone
RHETORIC Richard Toye
RISK Baruch Fischhoff and John Kadvany
RITUAL Barry Stephenson
RIVERS Nick Middleton
ROBOTICS Alan Winfield
ROCKS Jan Zalasiewicz
ROMAN BRITAIN Peter Salway

THE ROMAN EMPIRE Christopher Kelly
THE ROMAN REPUBLIC David M. Gwynn
ROMANTICISM Michael Ferber
ROUSSEAU Robert Wokler
RUSSELL A. C. Grayling
THE RUSSIAN ECONOMY Richard Connolly
RUSSIAN HISTORY Geoffrey Hosking
RUSSIAN LITERATURE Catriona Kelly
THE RUSSIAN REVOLUTION S. A. Smith
SAINTS Simon Yarrow
SAMURAI Michael Wert
SAVANNAS Peter A. Furley
SCEPTICISM Duncan Pritchard
SCHIZOPHRENIA Chris Frith and Eve Johnstone
SCHOPENHAUER Christopher Janaway
SCIENCE AND RELIGION Thomas Dixon
SCIENCE FICTION David Seed
THE SCIENTIFIC REVOLUTION Lawrence M. Principe
SCOTLAND Rab Houston
SECULARISM Andrew Copson
SEXUAL SELECTION Marlene Zuk and Leigh W. Simmons
SEXUALITY Véronique Mottier
WILLIAM SHAKESPEARE Stanley Wells
SHAKESPEARE'S COMEDIES Bart van Es
SHAKESPEARE'S SONNETS AND POEMS Jonathan F. S. Post
SHAKESPEARE'S TRAGEDIES Stanley Wells
GEORGE BERNARD SHAW Christopher Wixson
THE SHORT STORY Andrew Kahn
SIKHISM Eleanor Nesbitt
SILENT FILM Donna Kornhaber
THE SILK ROAD James A. Millward
SLANG Jonathon Green
SLEEP Steven W. Lockley and Russell G. Foster
SMELL Matthew Cobb
ADAM SMITH Christopher J. Berry
SOCIAL AND CULTURAL ANTHROPOLOGY John Monaghan and Peter Just
SOCIAL PSYCHOLOGY Richard J. Crisp
SOCIAL WORK Sally Holland and Jonathan Scourfield
SOCIALISM Michael Newman
SOCIOLINGUISTICS John Edwards
SOCIOLOGY Steve Bruce
SOCRATES C. C. W. Taylor
SOFT MATTER Tom McLeish
SOUND Mike Goldsmith
SOUTHEAST ASIA James R. Rush
THE SOVIET UNION Stephen Lovell
THE SPANISH CIVIL WAR Helen Graham
SPANISH LITERATURE Jo Labanyi
SPINOZA Roger Scruton
SPIRITUALITY Philip Sheldrake
SPORT Mike Cronin
STARS Andrew King
STATISTICS David J. Hand
STEM CELLS Jonathan Slack
STOICISM Brad Inwood
STRUCTURAL ENGINEERING David Blockley
STUART BRITAIN John Morrill
THE SUN Philip Judge
SUPERCONDUCTIVITY Stephen Blundell
SUPERSTITION Stuart Vyse
SYMMETRY Ian Stewart
SYNAESTHESIA Julia Simner
SYNTHETIC BIOLOGY Jamie A. Davies
SYSTEMS BIOLOGY Eberhard O. Voit
TAXATION Stephen Smith
TEETH Peter S. Ungar
TELESCOPES Geoff Cottrell
TERRORISM Charles Townshend
THEATRE Marvin Carlson
THEOLOGY David F. Ford
THINKING AND REASONING Jonathan St B. T. Evans
THOUGHT Tim Bayne
TIBETAN BUDDHISM Matthew T. Kapstein
TIDES David George Bowers and Emyr Martyn Roberts

TIME Jenann Ismael
TOCQUEVILLE Harvey C. Mansfield
LEO TOLSTOY Liza Knapp
TOPOLOGY Richard Earl
TRAGEDY Adrian Poole
TRANSLATION Matthew Reynolds
THE TREATY OF VERSAILLES Michael S. Neiberg
TRIGONOMETRY Glen Van Brummelen
THE TROJAN WAR Eric H. Cline
TRUST Katherine Hawley
THE TUDORS John Guy
TWENTIETHCENTURY BRITAIN Kenneth O. Morgan
TYPOGRAPHY Paul Luna
THE UNITED NATIONS Jussi M. Hanhimäki
UNIVERSITIES AND COLLEGES David Palfreyman and Paul Temple
THE U.S. CIVIL WAR Louis P. Masur
THE U.S. CONGRESS Donald A. Ritchie
THE U.S. CONSTITUTION David J. Bodenhamer
THE U.S. SUPREME COURT Linda Greenhouse
UTILITARIANISM Katarzyna de Lazari-Radek and Peter Singer
UTOPIANISM Lyman Tower Sargent
VETERINARY SCIENCE James Yeates
THE VIKINGS Julian D. Richards
THE VIRGIN MARY Mary Joan Winn Leith
THE VIRTUES Craig A. Boyd and Kevin Timpe
VIRUSES Dorothy H. Crawford
VOLCANOES Michael J. Branney and Jan Zalasiewicz
VOLTAIRE Nicholas Cronk
WAR AND RELIGION Jolyon Mitchell and Joshua Rey
WAR AND TECHNOLOGY Alex Roland
WATER John Finney
WAVES Mike Goldsmith
WEATHER Storm Dunlop
THE WELFARE STATE David Garland
WITCHCRAFT Malcolm Gaskill
WITTGENSTEIN A. C. Grayling
WORK Stephen Fineman
WORLD MUSIC Philip Bohlman
THE WORLD TRADE ORGANIZATION Amrita Narlikar
WORLD WAR II Gerhard L. Weinberg
WRITING AND SCRIPT Andrew Robinson
ZIONISM Michael Stanislawski
ÉMILE ZOLA Brian Nelson

Available soon:

VIOLENCE Philip Dwyer
JOHN STUART MILL Gregory Claeys
COGNITIVE BEHAVIOURAL THERAPY Freda McManus
HUMAN RESOURCE MANAGEMENT Adrian Wilkinson
ELIZABETH BISHOP Jonathan F. S. Post

For more information visit our website
www.oup.com/vsi/

Maciej Dunajski

GEOMETRY

A Very Short Introduction

OXFORD
UNIVERSITY PRESS

Great Clarendon Street, Oxford, OX2 6DP,
United Kingdom

Oxford University Press is a department of the University of Oxford.
It furthers the University's objective of excellence in research, scholarship,
and education by publishing worldwide. Oxford is a registered trade mark of
Oxford University Press in the UK and in certain other countries

First Edition published in 2022

Published in the United States of America by Oxford University Press
198 Madison Avenue, New York, NY 10016, United States of America

British Library Cataloguing in Publication Data
Data available

Library of Congress Control Number: 2021943454

ISBN 978-0-19-968368-0

Printed and bound by CPI Group (UK) Ltd, Croydon, CR0 4YY

Preface

Geometry is at least 2,500 years old, and it is within this field that the concept of mathematical proof—deductive reasoning from a set of axioms—first arose. Geometry is also a very active area of research in mathematics, and yet in the last few decades it has gradually disappeared from the high school syllabus in the UK and elsewhere. This book attempts to revitalize the subject.

I will proceed from concrete examples (of mathematical objects like Platonic solids or theorems like the Pythagorean theorem) to general principles. I assume little prior mathematical knowledge—a high school level is enough—but you should be willing to use a pencil, paper, compass, and ruler as you read. Geometry stands out from other branches of mathematics in that within this field a proof of a theorem can be given in pictorial terms without the need for algebra but at the same time without sacrificing rigour.

I have not used calculus, and tried to avoid too much algebra. The exception is Chapter 6, where matrices appear occasionally. The payoff will hopefully be high as you will be able to appreciate the link between modern geometry and symmetry, but this chapter is tangential to the rest of the book.

Chapter 2 is about Euclidean geometry and is best read before any of the later chapters, each of which focuses on a different modification of Euclid's postulates. Chapter 4, which introduces curved spaces, should be read before Chapter 7, where curvature appears in the context of Einstein's theory of gravitation. Chapter 5, which in many ways shows geometry at its best, will appeal to anyone wishing to enter geometry from the perspective of Renaissance art.

While writing this book I used my notes from popular geometry lectures that I gave to students applying to the University of Cambridge. I am grateful to Nick Woodhouse for his encouragement in the early stages of this project, and to Latha Menon and Robin Wilson for reading the earlier drafts and suggesting changes which resulted in several improvements. I am also grateful to Roger Penrose for sharing his geometric perspective on science with me, and for introducing me to classical projective geometry. Above all, I thank Adam Dunajski for carefully reading the manuscript, offering his criticism, and producing all the figures in the book.

Contents

List of illustrations xix

1 What is geometry? 1

2 Euclidean geometry 7

3 Non-Euclidean geometry 47

4 Geometry of curved spaces 66

5 Projective geometry 83

6 Other geometries 103

7 Geometry of the physical world 120

Further reading 143

Index 145

List of illustrations

1 The Pythagorean theorem **2**

2 A proof of the Pythagorean theorem **2**

3 Escher's *Circle Limit IV*. A model of hyperbolic geometry **3**
M.C. Escher's "Circle Limit IV" © 2021 The M.C. Escher Company-The Netherlands. All rights reserved. www.mcescher.com.

4 *The Last Supper*. All parallel lines intersect at one point **5**
Leonardo da Vinci/Wikipedia.

5 The Brianchon theorem **6**

6 π does not equal 3 **8**

7 Euclid's fifth axiom **10**

8 The parallel axiom **10**

9 Opposite angles **11**

10 Bisecting an angle **13**

11 Trisecting a line segment **13**

12 Two congruent polygons related by a sequence of isometries: a reflection with respect to line l, a rotation around point O, and a translation **17**

13 The 'British Rail' metric **18**

14 A convex and a non-convex polygon **23**

15 Sums of angles in the triangle **23**

16 Partitioning an n-gon **24**

17 Hexagonal tiling of a honeycomb **25**
Courtesy Dr Daniel Twitchen.

18 Tilings of the plane **26**

19 Penrose's aperiodic tiles, and their matching rules **27**

20 Five Platonic solids **29**

21 Finding the area of a triangle **32**

22 The Witch of Agnesi **36**

23 Cartesian coordinates of a point **37**

24 Vector addition **40**

25 Solving Sylvester's problem **44**

26 The chromatic number of the plane is at most 7 **46**

27 A great circle **48**

28 A spherical triangle **49**

29 Mercator's projection **51**

30 A lune **52**

31 Hyperbolic lines **54**

32 A construction of a hyperbolic line **55**

33 Infinitely many parallel lines **56**

34 The hyperbolic distance **56**

35 Hyperbolic and Euclidean distances from the centre of the disc **58**

36 *Circle Limit IV* (M. C. Escher) **59**

37 A hyperbolic triangle **60**

38 Tiling by ideal triangles **61**

39 Tiling with regular heptagons **62**

40 The tractrix and the pseudosphere **64**

41 The Beltrami model **65**

42 Extrinsic curvature of a curve **67**

43 Osculating circle **68**

44 Curvature of the parabola $y = x^2$ is $\varkappa = 2(1 + 4x^2)^{-3/2}$ **69**

45 Slicing a surface **70**

46 Gaussian curvature **70**

47 Genus 3 surface **74**

48 Stereographic projection. Two points, P and Q, are close to each other on the sphere, but their images, P' and Q', are far apart on the plane **76**

49 Tangent vector on the tangent plane **78**

50 Length of a curve 79

51 (a) Duccio di Buoninsegna (circa 1310), *The Capture of Jesus*. 84
Duccio di Buoninsegna/Wikipedia.

(b) Canaletto (circa 1730), *Interior Court of the Doge's Palace* 84
© Fitzwilliam Museum/Bridgeman Images.

52 A perspective machine, from Albrecht Dürer's *Painter's manual* (1525) 85
Bridgeman Images.

53 Projection from a line to a line 86

54 Projection from a plane to a plane 87

55 Desargues's theorem 88

56 Points and lines on a projective plane 90

57 Conic sections 93

58 The goat definition of the ellipse 93

59 Pascal's theorem 96

60 Poncelet's theorem 97

61 Projective duality 98

62 Duality with respect to a conic 98

63 Brianchon's theorem is the dual of Pascal's theorem 100

64 Ceva theorem 106

65 Transformations of a square in different geometries 111

66 Two quadruples of points with the same cross-ratio are related by a projective transformation 113

67 Fermat curves with $n = 2, 3$, and 4 115

68 Rational parametrization of a circle 116

69 Light cone in space-time 122

70 Geodesics are the longest paths 124

71 Simultaneity is not absolute 125

72 Distance in Aristotelian and Galilean space-times 129

73 Matter curving space-time 132

74 A black hole with the dashed vertical lines representing the event horizon 134

75 Gravitational collapse of a star to form a black hole 137

Chapter 1
What is geometry?

Sometime in the 6th century BC, the Greek philosopher Pythagoras of Samos and his followers, the Pythagoreans, spent time unveiling the relationship between numbers and geometric forms. They are credited with a proof of what is now known as the Pythagorean theorem: for any right-angled triangle, the square of the hypotenuse c is equal to the sum of the squares of the other two sides a and b (Figure 1).

The Pythagorean theorem is a result in *geometry*—from ancient Greek *geo*, meaning earth, and *metron*, measurement—a branch of mathematics concerned with lengths, shapes, and areas. Geometry stands out from most other branches of mathematics in that a proof of a theorem can be given in pictorial terms without the need for algebra or mathematical symbols but at the same time without sacrificing rigour (Figure 2).

While it may take some explaining to convince you that all the triangles in the left-hand square can be rearranged to those in the right-hand one, the proof is essentially self-contained. In fact, the concept of a mathematical proof—deductive reasoning from a set of *axioms*—first arose in geometry. Axioms are statements which are evidently true. They were first listed by another Greek mathematician, Euclid, in his *Elements*—possibly the most influential collection of mathematical books ever written.

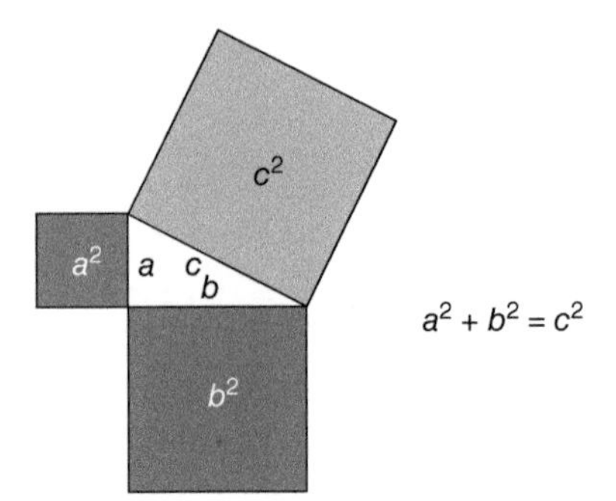

1. The Pythagorean theorem.

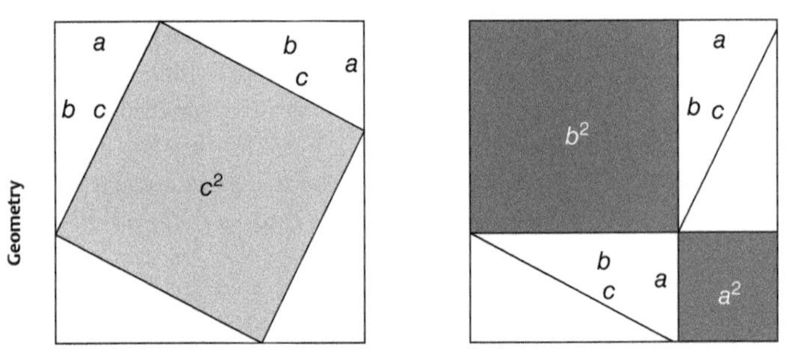

2. A proof of the Pythagorean theorem.

The Pythagorean theorem is valid in Euclidean geometry. It relies on a rather rich set of concepts: there are triangles and right angles, which both play a central role in the theorem. Then there are other polygons, such as squares, and each of these polygons has an area, which can be computed knowing only the sides and the internal angles of the polygon. Moreover, the theorem holds not only for right-angled triangles oriented horizontally, like the one in Figure 1, but also for triangles of any orientation on the plane. So it must be true that the concepts of angles and areas do not change under rotations and translations. While all this may be intuitively clear, even to a non-mathematician, it is a lot to take in. Which of the properties have to be assumed so that other properties can be deduced? Euclidean geometry takes care of all this, and this form of geometry will be the subject of Chapter 2.

Another theorem in Euclidean geometry is the statement that the sum of the angles in any triangle on a plane is 180°. Must this be true in other geometries? It turns out that the answer is no. The sum of the angles of a triangle on a sphere is always more than 180°. You may rightly object at this point, as it is not at all clear what a triangle on a sphere means. This will be explained in Chapter 3, where non-Euclidean geometries are introduced. Spherical geometry is one example of a non-Euclidean geometry, but there are others—historically the first being hyperbolic geometry—where one of Euclid's postulates is false. M. C. Escher's *Circle Limit IV* (Figure 3) represents this geometry on a disc with points on the boundary being infinitely far apart. In this geometry the analogues of straight lines are circular arcs intersecting the boundary of the disc at right angles. Six of these arcs have been drawn on Figure 3.

3. Escher's *Circle Limit IV*. A model of hyperbolic geometry.

Although the notion of distance is distorted in hyperbolic geometry, our everyday Euclidean notion of angles still makes sense. How about areas? It turns out that all the devils in Figure 3 have the same area. Those near the boundary seem smaller, but only because they are far away—with respect to the hyperbolic distance—from the centre of the disc. We have been fooled by our Euclidean intuition. We will come to this form of geometry in Chapter 3.

Spherical and hyperbolic geometries differ from Euclidean geometry in that *curvature* is present in both. Curvature is constant and positive for a sphere, and constant and negative for the hyperbolic disc. In Chapter 4 I shall introduce the mathematics needed to understand curved surfaces like the surface of a rabbit. Some regions of a rabbit look more curved than others, but it was only understood in the 19th century by the German mathematician Carl Friedrich Gauss that curvature is an intrinsic property of a surface, and depends only on measuring distances rather than on the way the surface sits embedded in the ambient space. This observation of Gauss has become known as the *Theorema Egregium* (Latin for 'Remarkable Theorem'). The Gaussian concept of curvature applies not only to surfaces but also to their higher-dimensional generalizations, known as *manifolds*. This leads to *differential geometry*—one of the most active research areas in mathematics—which we also explain in Chapter 4.

In Chapter 5 I shall introduce a very different type of geometry, where the notions of distance, angles, and areas do not exist. Instead, what matters is configurations of points and lines. The modern foundations of this kind of geometry were given by the 17th-century French mathematician Girard Desargues, and the geometry itself underlies the principles of perspective drawing, like that of Leonardo da Vinci, where all lines which would be parallel in ordinary three-dimensional space intersect at what we call *a point at infinity*. In *The Last Supper* (Figure 4), da Vinci has chosen this point to be in the right ear of Christ.

4. ***The Last Supper*****. All parallel lines intersect at one point.**

You may at this stage point out that we should not use the name geometry in its Greek meaning, for no measurements can be taken. The modern terminology for this set of ideas is *projective geometry*. The Pythagorean theorem cannot be true in projective geometry. For instance, take a photograph of a right-angled triangle, put it on a table, and look at it from 10 metres away. The right angle may appear obtuse or acute, depending on how you orient the triangle. The squares may also appear to be distorted and look like other quadrilaterals. Depending on the point from which the perspective is taken, the square of the hypotenuse can be either larger or smaller than the sum of squares of the remaining two sides. Are there, we may ask, any geometric theorems which remain valid in projective geometry? The answer is yes, and one example is the Brianchon theorem. It states that when a hexagon is circumscribed about an ellipse, the diagonals connecting opposite vertices meet at a single point (Figure 5). We shall meet a few more examples in Chapter 5.

Geometry has been one of the fastest-developing areas of mathematics, and it now overlaps with other branches of mathematics such as algebra, number theory, and topology. In fact, the proof of the celebrated Fermat's Last Theorem, given in 1995 by Sir Andrew Wiles, builds on results in *algebraic geometry*.

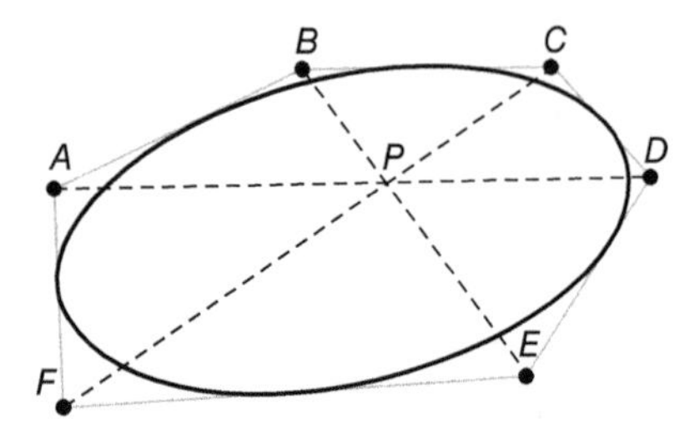

5. The Brianchon theorem.

In Chapter 6, we will take a short tour through other types of geometry not discussed in previous chapters.

One of the areas in which geometry plays a key role is physics. While Euclidean geometry gives a very accurate description of physical space, at large scales the Universe is described more accurately by a form of curved geometry that we will encounter in Chapter 4. According to Einstein's general theory of relativity, gravitation is just a manifestation of curvature. This theory works extraordinarily well, and its links to geometry will be explained in Chapter 7. The 2015 detection of gravitational waves—the ripples in the curvature of the Universe resulting from two black holes merging over 1.5 billion years ago—agrees with the predictions of the theory. This has been recognized by the award of a Nobel Prize in Physics in 2017 to Rainer Weiss, Barry Barish, and Kip Thorne. You may find it even more amusing that geometric ideas from Einstein's theory of gravitation also led to progress in pure mathematics. The proof of the Poincaré conjecture given by Grigori Perelman in 2003 is one illustration of the interaction between geometry and physics. Although a breakthrough in geometry, it captured the public imagination mainly because its author declined the Fields Medal awarded to him in 2006 and did not claim the $1 million reward offered by the Clay Mathematics Institute for solving one of the seven Millennium Prize Problems. As of 2021, the Poincaré conjecture is the only Millennium Prize Problem which has been solved.

Chapter 2
Euclidean geometry

Origins of geometry

The cultures that grew up in the arid region of Mesopotamia within the Tigris–Euphrates river system from the 4th millennium BC needed to tackle problems which we would now call geometrical. Dividing and surveying the land after periodic floods relied on measuring distances and computing areas. Development of trade required the Mesopotamian merchants to measure the quantity of grain in cylindrical or spherical containers.

The early methods varied from imperfect to precise. Areas of quadrilaterals were computed by multiplying the averages of the opposite sides. This gives the right answer for rectangles, but not for trapezia (quadrilaterals with only one pair of parallel sides). The Egyptians could calculate the volume of a square pyramid, like the Great Pyramid of Giza constructed around 2500 BC. They could also accurately calculate the volume of the cylinder as the product of its height and the area of its base. The rules of these computations were established empirically, and their correctness was measured by accuracy—the proofs were neglected. The geometrical statements were not meant to be absolutely exact, but only sufficiently good approximations. For example, the Mesopotamians took the value of π to be 3, although they most

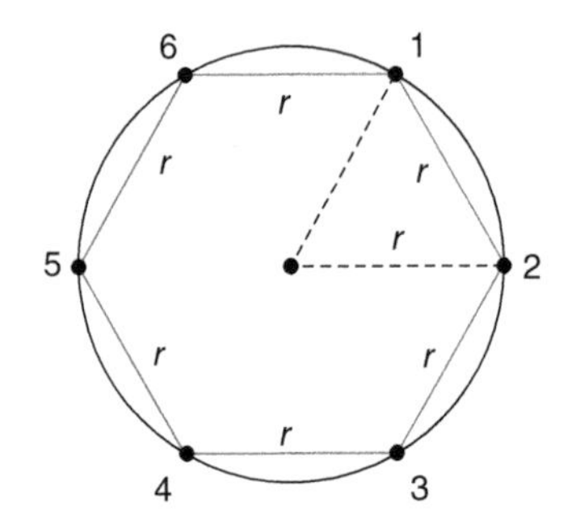

6. π does not equal 3.

likely realized that this was only an approximation. If you draw a circle and then place successive chords of length equal to that of the radius, then after six steps you will construct a hexagon and come back to the starting point (Figure 6). This gives the formula

$$\text{circumference} = 3 \times \text{diameter},$$

which cannot be right for a circle, as it is exactly right for the hexagon inscribed inside the circle. What, then, is the exact proportionality factor π between the circumference and the diameter of a circle? Does such a number even exist? There is no evidence that these questions occupied the Mesopotamians any more than they occupy the 21st-century school exam boards which quote π to be 3.14, or the manufacturers of eight-digit calculators which give 3.1415926.

While we know little about the origins of geometry, it is safe to assume that the Greeks learnt it, in its utilitarian form, from the Egyptians in around the 6th century BC. The Greek merchant and mathematician Thales of Miletus (640–546 BC) used the similarity of triangles to measure the height of the Great Pyramid of Giza by comparing the length of its shadow to the length of the shadow of a stick. He is credited with the first application of deductive methods.

The ancient Greek philosophers were not concerned with the practical applications of geometry, but rather with its relations to the absolute truth. They devised deductive and axiomatic methods to study the properties of volumes, areas, and distances. These methods have changed little in the past 2,500 years, so Greek geometry is where our story should begin.

Axioms and deduction

Euclidean geometry lies at the heart of geometry. It forms what we might call geometric intuition, and gives a very accurate description of the space around us. While the ancient Egyptians were familiar with geometry, they did not distinguish between abstract concepts and real objects. Thus a line segment connecting two points could have meant a piece of string between two pegs in the sand or a straight path between two villages. Other ancient cultures, such as the Chinese and Indians, engaged with geometry before the Greeks did, but primarily for practical purposes. In particular, Chinese mathematicians produced 'visual' proofs of geometric facts commonly attributed to the Greeks (like the Pythagorean theorem).

It was the Greeks who made geometry abstract. In his collection of books called the *Elements*, written in the 3rd century BC, Euclid of Alexandria put geometry into a logical framework. Starting from intuitively clear building blocks like points and lines, he made five assumptions which we now call *axioms* or *postulates*.

E1 Exactly one straight-line segment may be drawn between any two points.
E2 A straight-line segment may be extended indefinitely.
E3 A circle may be drawn with any given centre and an arbitrary radius.
E4 All right angles are equal.
E5 If a straight line crossing two line segments makes the interior angles on the same side less than two right angles,

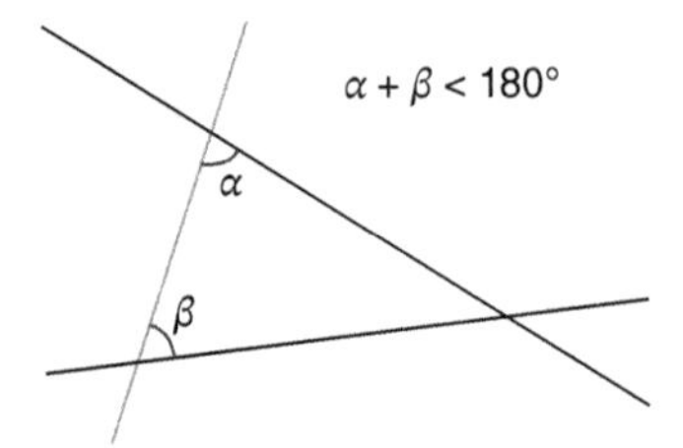

7. Euclid's fifth axiom.

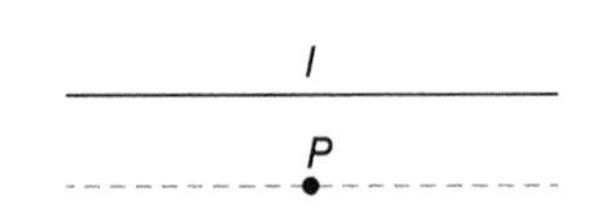

8. The parallel axiom.

the two segments, if extended indefinitely, intersect on the side on which the angles are less than the two right angles (Figure 7).

The first three axioms are intuitively clear. The fourth one looks a bit odd, but all it means is that given two pairs of lines such that the lines in each pair make a 90° angle with each other, one pair can be rotated and translated to coincide with the second pair. In this chapter and the next, we shall spend some time unveiling the meaning of the fifth axiom, which is equivalent to the following statement:

E5' (*The parallel postulate*) Given a line l and a point P not on l, there is exactly one line which contains P and does not intersect l (Figure 8).

Using these axioms, Euclid could define more complicated objects, like triangles. Finally, there came *theorems*, which are logical consequences of axioms. Some theorems took a long time to prove, while others were intuitively obvious, such as the statement that

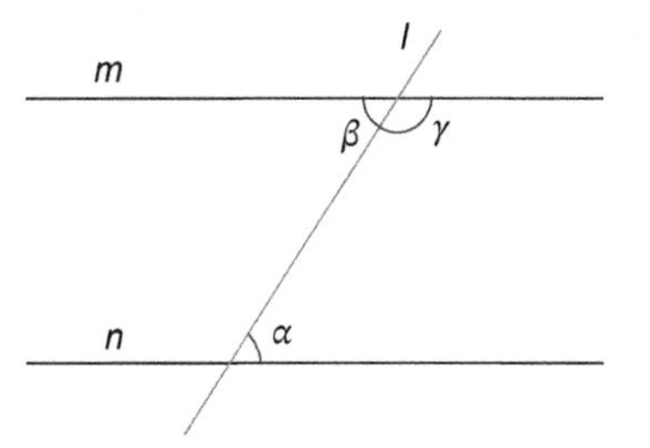

9. Opposite angles.

two distinct straight lines can meet at one point at most. A proof of this may be given by first assuming that the statement is false and then showing that such an assumption leads to a contradiction. So let us assume that the two lines meet at two points—call them P and Q. Then there would exist at least two lines through P and Q, which would, however, contradict the first axiom **E1**. Thus our assumption must have been false. This is an example of *proof by contradiction*—one of the finest weapons at a mathematician's disposal.

Another consequence of the axioms is that if a line l meets two parallel lines m and n, then the opposite angles are equal (Figure 9). If $\alpha + \gamma$ were bigger than 180°, the two lines m and n would intersect on the left of l; this is a consequence of **E5**. Similarly, if $\alpha + \gamma$ were smaller than 180°, then m and n would intersect on the right of l. Therefore, $\gamma = 180° - \alpha$. Moreover, $\beta + \gamma = 180°$, and so $\alpha = \beta$.

How can we be sure that some of Euclid's five axioms are not in fact theorems, which can be deduced from a smaller set of axioms? For over two millennia, some followers of Euclid had suspected that Euclid's fifth axiom—the parallel postulate—can be deduced from the first four. The matter was brought to a close in the 19th century by Nikolai Lobachevsky, Janos Bolyai, and Carl Friedrich Gauss, who gave an example of a geometry consistent with the first four axioms, but such that the parallel axiom does not hold. This geometry will be the subject of Chapter 3, and so in the rest of this chapter we shall focus on Euclidean geometry, where all five axioms are assumed to hold.

Compasses and ruler

The geometric shapes singled out by the axioms are line segments and circles. The tools reflecting this in geometric constructions are idealized rulers and compasses. Compasses consist of two rods joined together by a hinge which can be adjusted to any angle between 0° and 180°. One rod can be pinned into any point on the plane, and the other rod ends with a sharp pencil. Thus circles with arbitrary radii and centres can be drawn. A ruler can be used to draw a straight-line segment. It is not graded, meaning that no distances have been marked on the ruler—the Euclidean geometry has no preferred scale.

Let us consider a couple of examples of geometric constructions using a ruler and a compass, which are made possible by the axioms: bisecting an angle and trisecting a line segment. Given three points A, O, B, we connect A to O and O to B by two line segments, and denote the resulting angle by $\angle AOB$. To bisect this angle, place the point of the compass at the point O and draw an arc CD with point C lying on the line segment OA and point D on the line segment OB. Then draw two more arcs with the radius $|CD|$, one centred at C and the other at D. These two arcs intersect at two points. The segment connecting these two points can be extended to contain the point O. It divides the angle $\angle AOB$ into two equal angles (Figure 10).

The second construction trisects a line segment joining the points A and B. This problem involves a construction of a point P on the line segment AB such that $3|AP| = |AB|$. There are several ways to do this, using a combination of circles and lines. I shall use a

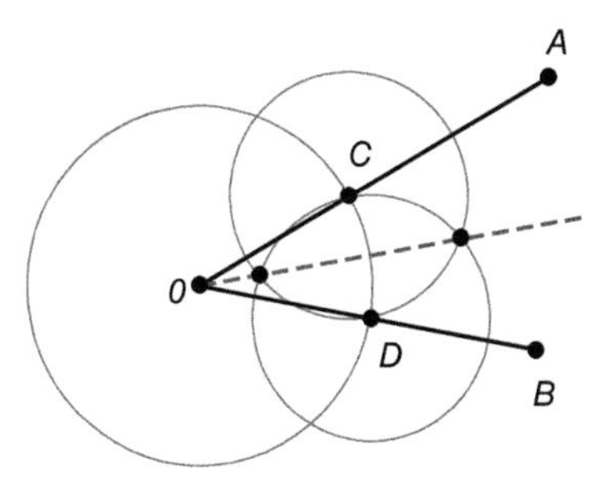

10. Bisecting an angle.

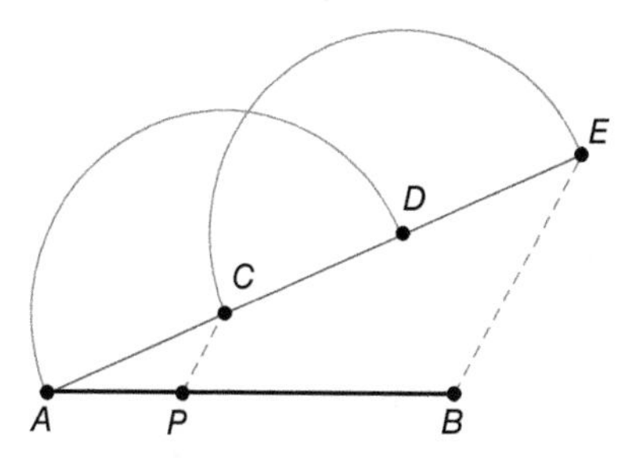

11. Trisecting a line segment.

method based on two circles, three lines, and the concept of similar triangles. This can be broken into six steps.

Step 1 Draw a line segment AC at an acute angle to the given segment AB, and use postulate **E2** to extend AC to a line l (Figure 11).

Step 2 Draw a circle with radius $|AC|$ and centre C. This circle intersects the line l containing the segment AC in the point D.

Step 3 Repeat Step 2: construct a circle with radius $|CD|$ and centre D. This circle intersects the line l at C and at another point which we shall call E. The point C trisects the segment AE, so that $|AE| = 3|AC|$.

Step 4 Connect the points E and B with a line segment. This gives the triangle with vertices A, B, and E.

Step 5 Construct the line parallel to BE and containing C. Let P be the intersection point of this line with the segment AB.

Step 6 The triangles AEB and ACP are similar; therefore,

$$\frac{|AC|}{|AE|} = \frac{|AP|}{|AB|}, \quad \text{so} \quad 3|AP| = |AB|.$$

Note that Steps 2 and 3 can be repeated any number of times. We can therefore divide a line segment into n parts, where n is any natural number. Or can we? The last step relies on the similarity of triangles and ratios, which we have not proved. It can, however, be deduced from Euclid's axioms.

These two constructions—bisecting an angle and trisecting a line segment—are not unrelated, as a line segment can be regarded as a limiting case of an angle with 180°. Thus, bisecting a line segment is a special case of bisecting an angle. This suggests that a hybrid construction which the Greeks have attempted—trisecting an angle—may also be possible using only a ruler and compasses. It turns out that it is not, but the proof of impossibility was only given in 1837 by the French mathematician Pierre Wantzel. Wantzel's methods used a mélange of algebra and number theory, both beyond the scope of this book. It is, however, a misconception to conclude that trisecting an angle is impossible. One cannot achieve it using circles and lines, but it can be done using other curves. I shall return to this point in Chapter 6.

Lengths and numbers

The angle scale is absolute in Euclidean geometry—axioms **E4** and **E5** single out right angles and use them to describe parallel lines. A natural unit of angular measure is the right angle. This can be divided or multiplied to obtain other angles, and the two standard units of angle measure are degrees and radians. The radians, although less commonly used than degrees, are preferred by mathematicians as more 'geometric': one radian is the angle subtended at the centre of a circle of radius r by an arc of length r. Radians and degrees are related by

$$180^\circ = \pi \text{ radians} = 2 \text{ right angles.}$$

On the other hand, there is no preferred choice for the distance scale. Numbers in Euclidean geometry are expressed in terms of ratios of lengths—for example, the diagonal of a square to its side or the circumference of a circle to its diameter—rather than directly in terms of lengths. In fact, the notion of a distance does not feature among Euclid's axioms. The choice of a distance scale corresponds to choosing a *distance ruler* and declaring it to be of unit length. Any other length can then be compared with the unit by rotating and translating the chosen distance ruler in space. A customary distance ruler of choice in North America is an inch. It can be compared with rulers used in continental Europe, called centimetres. An inch is longer, but not in any way more or less fundamental than a centimetre. A ruler preferred by astronomers is called a parsec, and is equal to approximately 3.086×10^{18} centimetres. The astronomers would measure the ratio of an inch to a centimetre to be around 2.54, and so would nuclear physicists, whose preferred ruler—the *Bohr radius* of the hydrogen atom—is 5.29×10^{-9} centimetres. This is because a ratio of distances is well defined in Euclidean geometry, in that it does not depend on the choice of a distance ruler.

To make this more precise, we pick any two distinct points O and R in space, and join them by a line l which extends indefinitely. A distance ruler is an identification of this line with the number line $\mathbb{R}$, such that the point O corresponds to the number 0 and point R corresponds to 1. Any other points P and Q on the line l correspond, by our choice of distance ruler, to real numbers p and q, which may also be negative depending on the ordering of the four points O, P, Q, R on the line l. The distance $d(P, Q)$ between P and Q is then defined to be

$$d(P, Q) = |p - q|,$$

where the symbol $|\ |$ denotes the absolute value: $|x| = x$ if $x \geq 0$ and $|x| = -x$ if $x < 0$, so $|-2| = |2| = 2$. In Euclidean geometry, the distance between P and Q is synonymous with the length of the interval PQ, which we shall denote by $|PQ|$. The notion of the distance resulting from any choice of a distance ruler satisfies

D1 $d(P, Q) = d(Q, P)$
D2 $d(P, Q) = 0$ if and only if $P = Q$
D3 $d(P, Q) + d(Q, S) \geq d(P, S)$.

This last property holds for any three points P, Q, S, not necessarily on the same line. There is no need to choose different distance rulers in different directions, as one distance ruler can be translated and rotated. The property **D3** is called the triangle inequality, as it is equivalent to the statement that the sum of the lengths of any two sides of a triangle is greater than the length of the remaining side.

In geometry a transformation is a process of mapping a plane or a space to itself. It can be applied to shapes or collections of points. In Chapter 6 we will discuss a more formal approach based on functions, but in what follows we shall only consider transformations that preserve distances. Such transformations are called isometries. Each isometry is a combination of rotations, translations, and reflections. Thus, if T is an isometry and $T(P)$ denotes the image of point P after the isometry has been applied, then

D4 $d(T(P), T(Q)) = d(P, Q)$.

Apart from distances, isometries preserve angles and map lines to lines and circles to circles. If two geometric shapes can

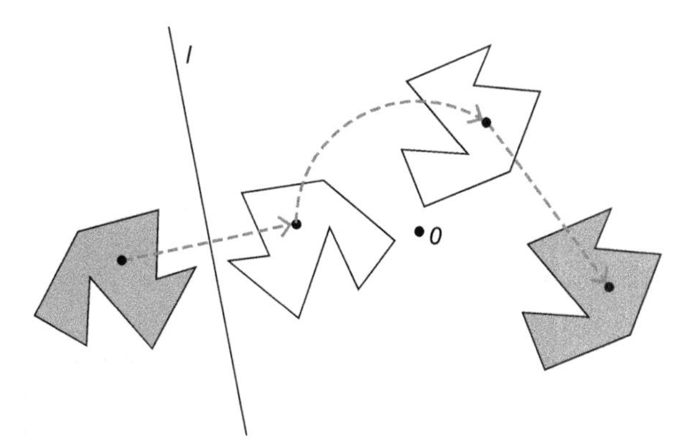

12. Two congruent polygons related by a sequence of isometries: a reflection with respect to line *l*, a rotation around point *O*, and a translation.

be transformed into each other by a combination of rotations, translations, and reflections, we call these shapes *congruent* (Figure 12).

I should stress that there are other ways to introduce distance in Euclidean geometry, which do not refer to choices of distance rulers. In *synthetic geometry* one defines $d(P, Q)$ by the properties **D1, D2, D3,** together with an assumption that a distance between any two points is nonnegative. These three properties alone do not single out the Euclidean distance we defined using the distance ruler. To give a different example of a distance on the plane consistent with **D1, D2, D3,** pick a point O and, for any other points P, Q on the plane, define

$$d(P, Q) = \begin{cases} |p - q| & \text{if } P, Q, \text{ and } O \quad \text{lie on the same line} \\ |p| + |q| & \text{otherwise,} \end{cases}$$

where the right-hand sides of this expression are defined using the Euclidean distance resulting from any choice of the distance ruler with the point O corresponding to 0. This notion of the distance is

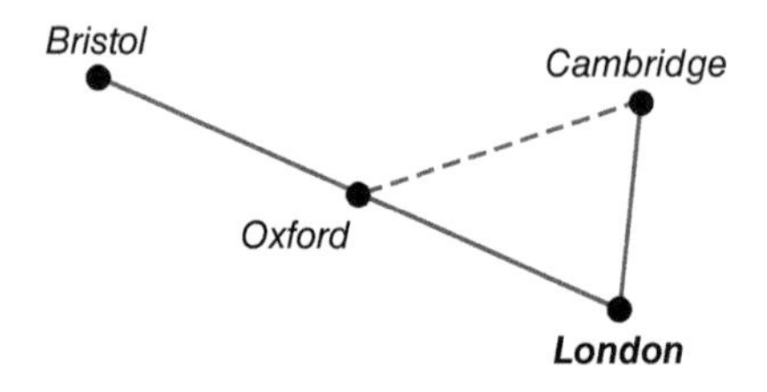

13. The 'British Rail' metric.

sometimes called the 'British Rail' metric, and the point O is referred to as London, from where all train lines radiate. Oxford and Bristol are on the same train line, so their British Rail distance equals the Euclidean distance. A train journey from Oxford to Cambridge is (at the time of writing) necessarily via London, which makes the British Rail distance between the two cities bigger than the Euclidean distance (Figure 13). You may want to verify that the British Rail metric does indeed satisfy the triangle inequality **D3** but not **D4**.

Let us return to the Euclidean distance defined by a choice of a ruler, and assume that a point P corresponds to a number p on an inch ruler used by Bob. Alice can also examine this point with her parsec ruler and find that it corresponds to a different number—call it p'. In fact, for any choice of P, the numbers p' and p are related by

$$p' = ap + b,$$

where a and b are some real numbers such that a is not equal to zero. The number b corresponds to the arbitrariness in the choice of origin, and a reflects the unit scales being different on the two rulers. In our case, a is around 8.23×10^{-19}. We say that p and p' are related by an *affine* transformation; I will have more to say about these transformations in Chapter 6. Bob now chooses four distinct points P, Q, U, V and computes the ratio of two intervals $|PQ| : |UV| = |p - q|/|u - v|$. Alice computes the same ratio using her ruler and finds

$$\frac{|PQ|}{|UV|} = \frac{|(ap+b)-(aq+b)|}{|(au+b)-(av+b)|} = \frac{|a||p-q|}{|a||u-v|} = \frac{|p-q|}{|u-v|}.$$

The ambiguity b of the choice of origin has cancelled in both the numerator and denominator, and the scale ambiguity a cancelled upon taking the quotient. Thus ratios of lengths do not depend on a choice of units, and so neither do the classical geometric constructions like trisecting the line segment I described earlier in the chapter. Alice and Bob would agree that New York is approximately twelve times further from Las Vegas than it is from Boston, but to have a constructive discussion about the actual distances, they should agree on the choice of ruler and fix the notion of a unit distance. Let us, from now on, assume that such a choice has been made.

Think of two line segments of different lengths. The segments are called commensurable if there exists a line segment of length r, such that both line segments can be partitioned into whole multiples of r. If no such r exists, then the segments are non-commensurable. An example of a non-commensurable segment is a side of a square of length d and a diagonal of this square, which, by the Pythagorean theorem, has length equal to $d\sqrt{2}$. The non-commensurability can be proved by contradiction: if the side and the diagonal were commensurable, then there would exist whole numbers m and n such that $m \times r \times d = n \times r \times \sqrt{2} \times d$, or

$$m^2 = 2n^2.$$

We can assume that m and n are the smallest such whole numbers, which therefore have no common divisors. Moreover, m^2 has to be even, which implies that m is also even, and equal to $2k$ for some whole number k. Substituting this into the relation above gives $2k^2 = n^2$, which can only hold if n is even, which contradicts our assumption.

The conclusion that no m and n can exist led to difficulties for the Pythagoreans, who used whole numbers, which we shall also call integers, like $1, 2, 7, 44, \ldots$, as well as rational numbers like $2/7$, or m/n where m and n are both integers, and $n \neq 0$. How, then, can one account for a number x which corresponds to the length of the diagonal of the unit square? By the Pythagorean theorem, this number must satisfy

$$x^2 = 1 + 1 = 2,$$

but the Pythagoreans could prove, perhaps using the argument we gave above, that no integers m and n can exist such that $m^2/n^2 = 2$. Using modern terminology, the value of x, $\sqrt{2}$ is an irrational number.

While this must have been a blow to the Pythagoreans, the square root of two was still accepted as a length. Moreover, although the intervals with lengths 1 and $\sqrt{2}$ are not commensurable, the length $\sqrt{2}$ relative to a unit distance can be constructed by using a compass and ruler, as it is possible to construct a square with any given side. It turns out—although it was only proved in the 19th century—that it is not possible to give such a construction for the cube root of two. This problem, known as the duplication of the cube, was one of the three problems of antiquity (the other two were trisecting an angle and constructing a square with the same area as that of a given circle). Legend has it that the oracle at Delphi ordered the size of the cube-shaped altar dedicated to Apollo to be doubled. Plato was consulted, but could not solve the problem using a ruler and compasses. If we declare the side of the original cube to be 1, then the doubled cube will have side x, where

$$x^3 = 2 \cdot 1^3$$

or $x = \sqrt[3]{2}$. The argument we used to prove that $\sqrt{2}$ is irrational can also be used to establish the irrationality of $\sqrt[3]{2}$, and therefore its non-commensurability with the side of the cube. This alone does not rule out a compass and ruler construction, as, after all, that is possible for $\sqrt{2}$. To see what is going on, let us examine more closely the algebraic operations involved in constructing commensurable segments. We shall say that a positive number x is *constructible* if and only if, given a line segment of unit length, a line segment of length x can be constructed by a finite iteration of ruler and compass constructions. In this way we can construct points as intersections of either two lines, a line and a circle, or two circles, assuming that the lines and circles have already been determined from the initial unit line segment. The number 1 is constructible, and so is any other number which can be obtained by applying square roots and four arithmetic operations, $+, -, \times, \div$, to constructible numbers. Thus the numbers

$$\sqrt{2}, \quad 3\frac{2}{13}, \quad \sqrt{17 + \frac{1}{3}\sqrt{11 + \sqrt{5}}} - 1$$

are all constructible. Let us focus on the last one of these and call it x. By reversing the allowed operations (first add 1 to x, square the answer, subtract 17, etc.), we find that x is a root of a polynomial equation of degree eight:

$$\left((3((x+1)^2 - 17))^2 - 11\right)^2 - 5 = 0.$$

In general, constructible numbers are roots of polynomial equations with integer coefficients whose degree—the highest power of x appearing in the equation—is equal to some power of two (this takes care of reversing all operations involving square roots). Moreover, these equations should be irreducible in the sense that they do not factor into lower-degree polynomials with integer coefficients. Therefore, $\sqrt[3]{2}$ is not constructible, as the polynomial equation

$$x^3 - 2 = 0$$

is of degree three, and three is not a power of two. The number $\sqrt[3]{2}$ also satisfies a degree-four equation,

$$x^4 + x^3 - 2x - 2 = 0$$

but this equation can be rewritten as $(x^3 - 2)(x + 1) = 0$, and so it is reducible.

Is the circumference of a circle commensurable with its diameter? Let us use the symbol π, which is the ratio of the circumference to the diameter. It turns out that neither π nor $\sqrt{\pi}$ arises as a root of any polynomial equation with integer coefficient. Such numbers are *transcendental*, and none of them are commensurable with the unit. The transcendence in particular implies that π is irrational, but even demonstrating the irrationality of π alone is far more difficult than the argument we gave for $\sqrt{2}$. This also proves that squaring the circle is impossible with a ruler and compasses construction. Say that the radius of the circle is 1, so that its area is π. The side of a square with the same area is $\sqrt{\pi}$, but this number is also transcendental; this follows from the transcendence of π.

The number π can, however, be approximated by rational numbers—the value 22/7 already gives a result correct to two decimal places. Archimedes of Syracuse sandwiched a circle between two polygons, one inscribed inside a circle and one circumscribed about a circle. Calculating both perimeters gives an upper and a lower bound for π. Using polygons with 96 sides leads to

$$3\frac{10}{71} < \pi < 3\frac{1}{7}.$$

Taking the average of the upper and the lower bound gives $\pi = 3.14185$, which is correct up to three decimal places.

Polygons and tilings: from Plato to Penrose

A polygon is a plane shape bounded by segments of straight lines. It is called convex if any two points inside the polygon can be connected by a line segment contained inside the polygon (Figure 14).

The simplest polygon is a triangle. I shall use the Euclid axioms to prove that the sum of its internal angles is 180°.

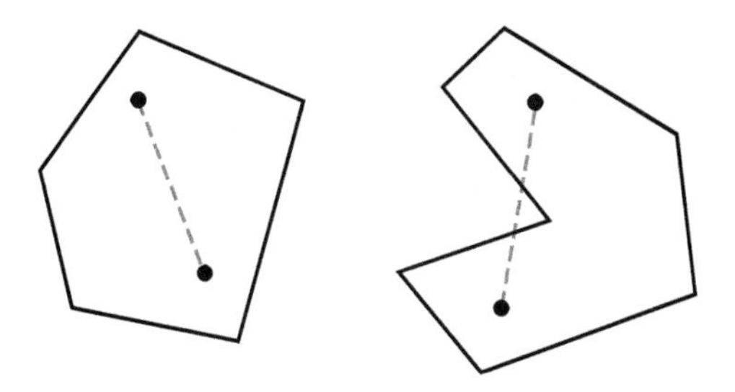

14. A convex and a non-convex polygon.

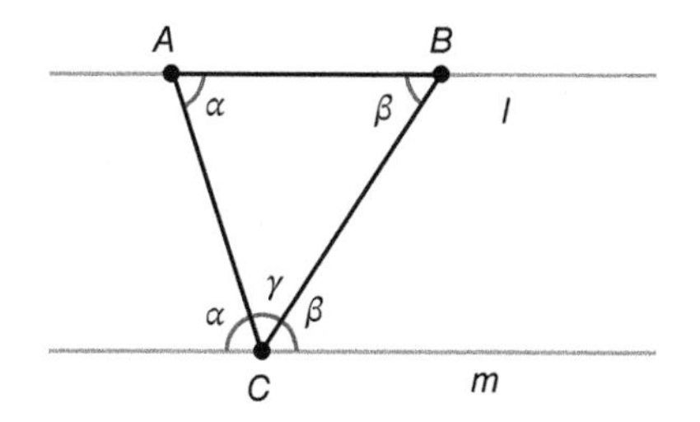

15. Sums of angles in the triangle.

Let α, β, γ be the angles emanating from the three vertices A, B, C (Figure 15). The axiom **E2** allows us to extend the segment AB to a line l. The fifth axiom in the form **E5'** implies that there exists the unique line m, which goes through C and is parallel to l. Using the properties of the opposite angles which we have already established from the axioms, we find

$$\alpha + \beta + \gamma = 180°.$$

Any convex polygon with n sides (we shall call such polygons n-gons) can be partitioned into $(n - 2)$ triangles (Figure 16). The sum of the angles in any of the triangles is 180°, which gives the sum of all internal angles of the convex n-gon to be $(n - 2) \times 180°$.

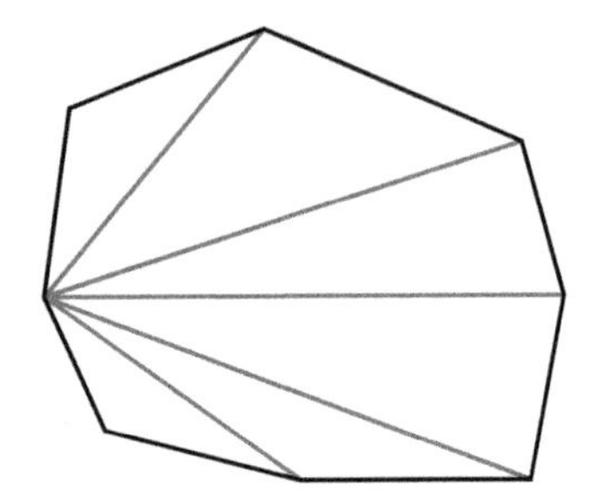

16. Partitioning an n-gon.

A polygon is called regular if all its segments have equal length and meet at equal angles. Thus an equilateral triangle is a regular 3-gon, and a square is a regular 4-gon. The next three are pentagon, hexagon, and heptagon. All internal angles are equal in a regular polygon, and there are n of them. Thus an internal angle is the total sum of all angles divided by n, which gives the expression for an internal angle

$$\alpha = 180^\circ - \frac{360^\circ}{n}.$$

An equilateral triangle, a square, and a regular pentagon can all be constructed using a ruler and compasses, by inscribing them in a circle. Moreover, if a regular polygon can be constructed in that way, then so can a polygon with twice the number of sides. It is not possible to construct all regular polygons with the aid of a ruler and compass. This, together with a criterion for the n-gons to be constructible, was established by the German mathematician Carl Friedrich Gauss when he was 19 years old. A regular polygon can be constructed with compass and ruler if and only if the number of its sides is a product of a power of two and any number of different prime numbers of the form

$$2^{2^m} + 1,$$

where m is a natural number. The first five prime numbers of this form are $3, 5, 17, 257, 65537$. Thus, for example, a regular 9-gon

17. Hexagonal tiling of a honeycomb.

cannot be constructed with compass and ruler, as $9 = 3 \times 3$ (so that $m = 0$) but three is not a power of two.

Let us examine a honeycomb made by bees (Figure 17). The regular pattern of the honeycomb can be repeated indefinitely, to cover the whole plane with regular hexagons with no gaps or overlaps. A pattern with this property is called a tiling. Circles could be used instead, but would leave gaps between them; this would not be a tiling. The bees could have also chosen a tiling with squares or equilateral triangles, but the hexagonal tiling uses the least amount of material to create perimeters between cells of a given area. In this way, more honey can fit in. Aren't the bees clever?

Apart from using regular hexagons, squares, and equilateral triangles, it is not possible to tile a plane with any other regular n-gons of one type. To tile the plane, the internal angle α of a regular polygon must divide $360°$, as otherwise it would not be

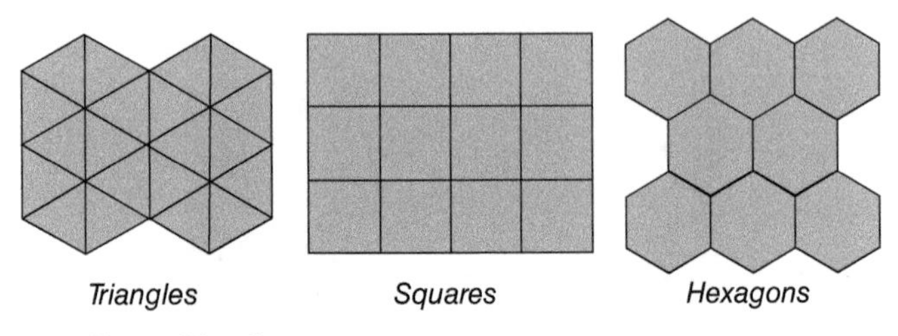

18. Tilings of the plane.

possible for a whole number of the polygons to meet at a point and leave no gaps. The equation $k\alpha = 360°$ has only three solutions with integers n and k: ($n = 3, k = 6$: triangles), ($n = 4, k = 4$: squares), and ($n = 6, k = 3$: hexagons) (Figure 18).

More tilings are possible if we relax some of the conditions. Tilting and stretching the square tiling gives a tiling with identical parallelograms. Allowing more than one type of regular polygons leads to infinitely many possibilities, some of which were used by medieval Islamic artists in floor or wall coverings. These tilings are all periodic in the sense that translating all tiles in two non-parallel directions by a fixed distance gives a tiling identical to the original. For example, the square tiling from Figure 18 can be translated up, down, left, or right by a distance equal to the side of a square, or diagonally by $\sqrt{2}$ times the side length.

A plane tiling with no period and such that it does not contain arbitrarily large periodic parts is called *aperiodic*. Such a tiling cannot be created by taking one of its sections and repeating it. There is some evidence that aperiodic tilings were known to Islamic artists, and an example of such a tiling can be found on the Darb-e Imam shrine in Isfahan, Iran. In 1974 the British mathematician and theoretical physicist Roger Penrose came up with several systematic methods of covering an entire plane in an aperiodic way. One of these involves two shapes of tiles—*fat and thin rhombi*—with equal sides but different angles (Figure 19).

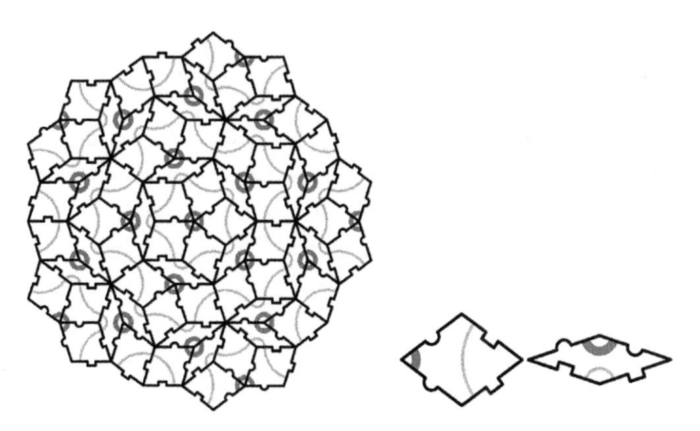

19. Penrose's aperiodic tiles, and their matching rules.

The internal angles are 72° and 108° in the fat rhombi, and 36° and 144° in the thin rhombi. Both the fat and thin rhombi will separately tile the plane, so the two types of tiles must be supplemented by matching rules which guarantee that the resulting tiling is aperiodic. Where this rule for the rhombi is presented (Figure 19), the similarly shaded parts are matched against each other. Equipped with a large set of tiles of both types, you now try to arrange them following the rules and cover a larger and larger region. You can, nevertheless, get stuck, and reach a stage where no more tiles can be added. Then you need to backtrack and make different choices along the way. The choices leading to a tiling are not unique, and there are infinitely many patterns built from both rhombi. As the tiling is expanded to cover a larger area, the ratio of the quantities of the fat rhombi to the quantities of the thin rhombi approaches the golden ratio,

$$\frac{1+\sqrt{5}}{2} = 1.618\ldots$$

This can be turned into a proof that this tiling is indeed non-periodic, because for any periodic tiling the ratio would be a rational number (as the ratio of some finite number of fat and thin rhombi), and the golden ratio is irrational.

Another property of Penrose's aperiodic tilings is their non-locality. Imagine that, having built a patch covering a large portion of the plane, you need to make two choices for how to place two tiles at opposite ends of the patch which may be billions of units away from each other, where the side length of any rhombi is one. It turns out that the choice you make at one end of the patch uniquely determines the type of tile which needs to be placed at the other end. Thus there is no local algorithm which, if followed, will lead to a tiling.

Aperiodic tilings occur in nature in the form of quasi-crystals, whose existence was conjectured by Penrose. In the mid 1980s, non-periodic atomic structures of crystals were indeed discovered by Dan Shechtman, who was awarded the Nobel Prize in Chemistry for this in 2011. It is still not known whether there exists an aperiodic tiling of a plane with only one type of tile.

Platonic solids

Analogues of polygons, but in space, are called polyhedra. A polyhedron is bounded by planes forming the faces. The meeting of two faces is called an edge, and edges intersect at a vertex. We shall assume that all polyhedra are convex, meaning that a segment joining any pair of points inside a polyhedron lies entirely within that polyhedron. A polyhedron is called regular, or a Platonic solid, if all its faces are identical regular polygons. Thus the same number of edges meets at each vertex, and the angles between any two edges are the same. The names of the Platonic solids reflect the Greek words for the number of faces. A tetrahedron has four faces which are equilateral triangles; a hexahedron (commonly known as a cube) has six faces which are squares; an octahedron has eight faces which are equilateral triangles; a dodecahedron has twelve faces which are regular pentagons; and an icosahedron has twenty faces which are equilateral triangles (Figure 20).

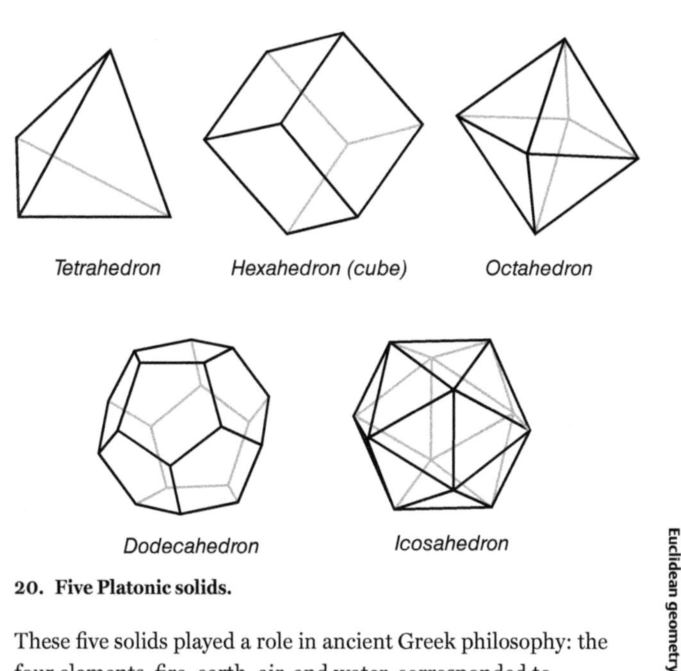

20. Five Platonic solids.

These five solids played a role in ancient Greek philosophy: the four elements, fire, earth, air, and water, corresponded to tetrahedron, cube, octahedron, and icosahedron, respectively. The dodecahedron represented the heavenly firmament.

There are no other Platonic solids apart from the five in Figure 20. To prove this, suppose that f is the number of faces meeting at each vertex, and that each face is that of a regular n-gon. Both f and n are clearly greater than or equal to three. The sum of angles at a vertex of a polyhedron is less than 360°, as unfolding a vertex and its adjacent faces onto the plane necessarily leaves a gap. Therefore, using our formula for the internal angles of a polygon gives

$$\left(180^\circ - \frac{360^\circ}{n}\right) \times f < 360^\circ$$

which, with some algebra, can be rearranged into the inequality

$$(n-2) \times (f-2) < 4.$$

It follows that $(n-2)$ and $(f-2)$ are positive integers whose product is smaller than four. There are only five such pairs of integers,

$$(1,1), \quad (1,2), \quad (2,1), \quad (1,3), \quad (3,1),$$

which gives

$n=3,$	$f=3$	tetrahedron	$V=4,$	$E=6,$	$F=4$
$n=4,$	$f=3$	cube	$V=8,$	$E=12,$	$F=6$
$n=3,$	$f=4$	octahedron	$V=6,$	$E=12,$	$F=8$
$n=5,$	$f=3$	dodecahedron	$V=20,$	$E=30,$	$F=12$
$n=3,$	$f=5$	icosahedron	$V=12,$	$E=30,$	$F=20.$

In the last three columns I have listed the numbers V, E, F of vertices, edges, and faces. These values are different for each Platonic solid, but in each case they are related by the formula

$$V-E+F=2.$$

This formula remains true for any polyhedron. It was discovered by the 18th-century Swiss mathematician Leonhard Euler, and the formula now carries Euler's name. To see why it is true, imagine inscribing a cube inside a sphere and continuously pushing the faces of the cube onto the surface of the sphere in a way which preserves the number of faces, edges, and vertices. In this way, the cube is turned into six regions on the sphere, divided by twelve boundaries, meeting at eight points. Each polyhedron can be given such a spherical representation. We can aim to construct an arbitrary polyhedron by drawing a polygon on the sphere consisting of intersecting arc and adding more edges to it. At the first step, when there is only one such polygon, we have $F=2$ (one face is the polygon itself, and the other one is its complement on the sphere) and $V=E$, as for one polygon the number of vertices equals the number of edges. Therefore, $V-E+F=2$, and the Euler formula holds. Attaching additional edges can then be accomplished in two ways.

1 Add a new vertex, and draw an edge joining it to one of the existing vertices.

This increases both V and E by 1 and does not change F, so the Euler formula remains valid.

2 Add a new edge and connect it to two already existing vertices.

This increases F and E by 1 and does not change V, and the Euler formula is not affected. As a projection of any polyhedron onto a sphere can be reconstructed by successive applications of **1** and **2**, the Euler formula holds for all polyhedra.

A remarkable symmetry relates different Platonic solids to each other. Take a Platonic solid **P**, and join the midpoints of all adjacent faces by edges. This gives a new polyhedron inscribed in **P** called the dual of **P**, which is also Platonic. This duality interchanges the numbers of vertices and faces, and leaves the number of edges intact. Moreover, if the process is repeated twice, the original Platonic solid is recovered. The dual of the icosahedron is the dodecahedron, and the dual of the octahedron is the cube. The tetrahedron is self-dual, as the duality operation produces another tetrahedron.

The Platonic solids occur in nature: the tetrahedron, cube, and octahedron appear in crystal structures, and all five appear as shapes of some micro-organisms and viruses. There are molecules of carbon which are polyhedra with 20 hexagonal and 12 pentagonal faces, 60 vertices, and 90 edges. This configuration, which satisfies the Euler formula, is called a fullerene and is named after the 20th-century American architect Richard Buckminster Fuller, who designed dome structures in which the hexagons and pentagons are further divided into triangles.

Areas

Our intuitive concept of the area of a geometric shape as the amount of paint needed to cover the shape is a good one, as long as

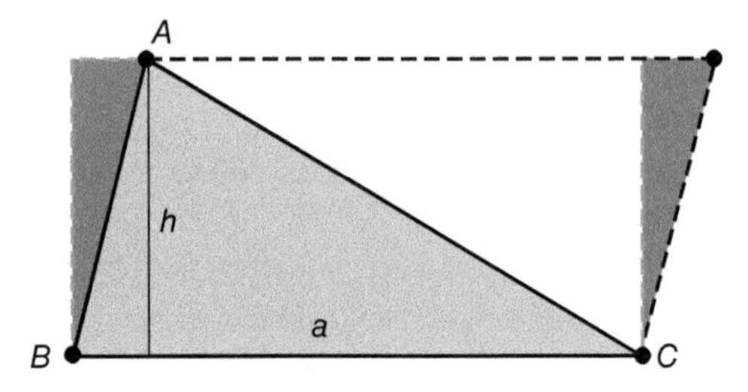

21. Finding the area of a triangle.

we exclude some pathological cases. This task turns out to be difficult, and has evolved into a separate branch of mathematics called *measure theory*. I shall first consider only polygonal shapes which can be bounded by a finite collection of line segments. The area of a polygonal shape $\mathcal{S}$ is a nonnegative real number called Area($\mathcal{S}$) which satisfies three axioms:

A1 If $\mathcal{S}$ and $\mathcal{U}$ are two polygons with no points in common, then

$$\text{Area}(\mathcal{S} \cup \mathcal{U}) = \text{Area}(\mathcal{S}) + \text{Area}(\mathcal{U}),$$

where $\cup$ denotes a union of two sets.

A2 The area of a rectangle with sides a and b is $a \cdot b$.

A3 Any congruent polygons have the same area.

These axioms define area by its properties—not an uncommon approach in mathematics. They lead to constructive methods to compute areas. For example, let us compute the area of a triangle with vertices ABC. Rotate the triangle by 180° about the midpoint of the segment AC. This gives a parallelogram whose area is equal to twice the area of the triangle; here we are using **A3** followed by **A1**. This area is equal to that of a rectangle with base BC and height h—to see it shift the right-angled triangles (Figure 21), and again refer to **A1**. Finally, **A2** gives

$$\text{Area}(\Delta ABC) = \frac{1}{2} a \times h.$$

We can also deduce from the axioms that if a polygon $\mathcal{S}$ is contained in a polygon $\mathcal{U}$, then the area of $\mathcal{S}$ is not greater than the area of $\mathcal{U}$. To see this, write

$$\mathcal{U} = \mathcal{S} \cup (\mathcal{U} \setminus \mathcal{S}),$$

where $\mathcal{U} \setminus \mathcal{S}$ is the complement of the set $\mathcal{S}$ in the set $\mathcal{U}$. It consists of all elements of $\mathcal{U}$ which are not elements of $\mathcal{S}$. The two sets $\mathcal{S}$ and $\mathcal{U} \setminus \mathcal{S}$ have no points in common, so **A1** gives

$$\text{Area}(\mathcal{U}) = \text{Area}(\mathcal{S}) + \text{Area}(\mathcal{U} \setminus \mathcal{S}) \geq \text{Area}(\mathcal{S}),$$

as Area$(\mathcal{U} \setminus \mathcal{S})$ is nonnegative by definition.

The property we have just established allows us to prove that the area of a line segment is zero. Indeed, a line segment of length a can be enclosed inside a rectangle with sides of lengths $a + \epsilon$ and ϵ, where ϵ is any positive number which we can take to be as small as we like. Therefore the area of the line segment is not greater than $(a + \epsilon) \cdot \epsilon$, by **A2**. But this number can be made arbitrarily small by taking the limit $\epsilon \to 0$.

To appreciate the need to exclude the pathologies I alluded to before formulating the axioms, let us consider two shapes which we shall call $\mathcal{R}$ and $\mathcal{I}$. The shape $\mathcal{R}$ consists of all straight-line segments which go through the centre of a unit square, have endpoints on the opposite edges of this square, and form angles which are rational numbers with the horizontal line through the centre of the square. The shape $\mathcal{I}$ consists of all such line segments, where the angles are irrational. We shall, for convenience, declare that the centre of the square belongs to $\mathcal{R}$ and not $\mathcal{I}$. In this way, any point on the square belongs to either $\mathcal{I}$ or $\mathcal{R}$, so that the square is the union of $\mathcal{I}$ and $\mathcal{R}$. What, then, is the area of $\mathcal{R}$? Any line segment in $\mathcal{I}$ is arbitrarily close to a line segment in $\mathcal{R}$, as an irrational angle can be approximated by a rational angle with an arbitrary precision, for example by truncating the decimal expansion of the angle. If we use red paint to colour $\mathcal{R}$, the whole square will look red. Is the area of $\mathcal{R}$ equal to one? Or is it zero, as we have already shown that any line segment has zero area? Or is it perhaps a half? The regions $\mathcal{I}$ and $\mathcal{R}$ are not polygonal, so we should give up hope of assigning an area to either of them.

We should be worried that by trying to exclude the anomalies like $\mathcal{R}$ and $\mathcal{I}$ above, we have made it impossible to define areas for shapes like circles and other regions enclosed by curves. A circle is not a polygon, but it can be approximated by one with arbitrary precision. In the first approximation, consider a circle of radius r inscribed in a square of side $2r$. The area of this square is $4r^2$, so if a circle has an area which can be made sense of, it should be less than that. We instead take a square inscribed in the same circle. This smaller square has a diagonal of length $2r$ and a side of length $r\sqrt{2}$, so its area is $2r^2$. Thus the area of a circle with radius r should be a number between $2r^2$ and $4r^2$. This crude approximation can be improved by sandwiching a circle between two regular n-gons—one inscribed inside a circle and one circumscribed about a circle—and obtaining lower and upper bounds for the area in that way. As n gets bigger, the approximation improves. In the limit where n tends to ∞, the area of the region between two polygons tends to zero and the area of both polygons tends to the area of the circle. A regular n-gon inscribed inside a circle of radius r consists of n isosceles triangles, each with base d_n. In the limit when n is large, the total area of these triangles tends to the area A of the circle, so that

$$n \times \left(\frac{1}{2} r \times d_n\right) \longrightarrow A, \quad \text{when} \quad n \to \infty.$$

The circumference $C = 2\pi r$ of the circle is the limit of nd_n as n tends to infinity, so the two formulae together give

$$A = \frac{1}{2} rC, \quad \text{or} \quad A = \pi r^2.$$

To make this procedure more formal, and to allow for shapes enclosed by other smooth curves, we can go back to squares and rectangles. Draw such a shape $\mathcal{S}$ on graph paper, and count the number of squares contained inside the shape. This gives a lower bound for the area. Including the squares which enclose the perimeter of the shape gives the upper bound. As the grid of the graph paper gets smaller, these upper and lower bounds tend to limits. If these limits exist, and are equal, then they define the area of $\mathcal{S}$ which, by its construction, is nonnegative and satisfies the axioms **A1, A2, A3**. For general shapes, the upper and lower bounds may still tend to limits which may, however, be different. In this case, no notion of area which satisfies **A1, A2, A3** can be assigned to $\mathcal{S}$.

Cartesian coordinates

The advancement of Euclidean geometry has revealed properties of geometric shapes such as cones, circles, and Platonic solids. The methodology was to examine these shapes and their areas, interior angles, and distances between vertices. Until the 17th century, however, it was not clear, what a geometric shape was. The shapes were defined by their properties. For example, a sphere consists of all points which are equidistant from a fixed point called the centre. A triangle is a polygon with three edges and three vertices. More exotic shapes were given catchy names, like the curve known as the 'Witch of Agnesi' studied by the 18th-century Italian mathematician Maria Agnesi.

To construct the Witch, consider two points O and B on the plane, and draw a circle with diameter OB. For any point C on this circle, let D be the point of intersection of the secant line OC and the tangent line to the circle at B. Now draw a perpendicular to the diameter OB through the point C. Let P be the point of intersection of this perpendicular with the line parallel to OB through D. Different choices of points C on the circle will lead to different Ps, and the locus of all points P is the Witch of Agnesi (Figure 22).

There were many more shapes with similarly beautiful geometric constructions, but a modern definition of a geometric shape—as a

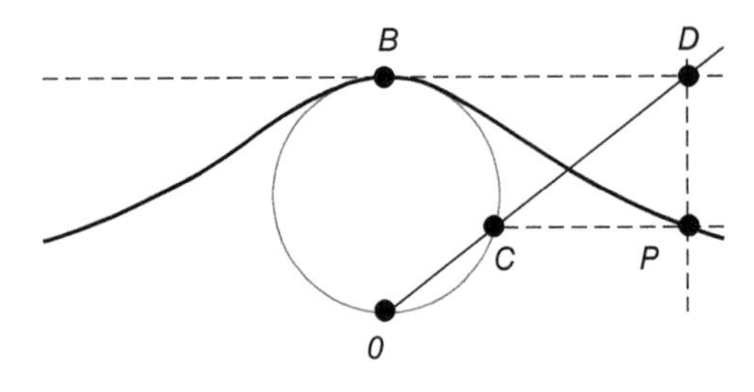

22. The Witch of Agnesi.

set of points on the plane or in space—could not have been formulated until the end of the 19th century, when the set theory was developed. Geometry was in crisis, as the precise subject of its research was blurred. The resolution was provided by the French philosopher and mathematician René Descartes, also known as Cartesius. In his *La Géometrié*, published in 1637, he married geometry with algebra. This eventually gave rise to a universal description of all geometric shapes by equations. The idea was to introduce a coordinate system: three pairwise perpendicular oriented straight lines l_1, l_2, and l_3 intersecting at one point called the origin O. Any point P in space is then uniquely described by three ordered numbers (p_1, p_2, p_3) called coordinates, where the p_1 coordinate is the distance, taken with the appropriate sign, of a perpendicular projection from P to l_1, with analogous constructions of p_2 and p_3. Conversely, specifying three coordinates uniquely determines a point in space (Figure 23).

If Q is another point with coordinates (q_1, q_2, q_3), then the distance between P and Q is given by the formula

$$d(P, Q) = \sqrt{(p_1 - q_1)^2 + (p_2 - q_2)^2 + (p_3 - q_3)^2}.$$

This is the Pythagorean theorem expressed in the Cartesian coordinate system. Geometric shapes can then be defined as algebraic relations between the coordinates of their points. For example, a sphere of radius r centred at the origin O with

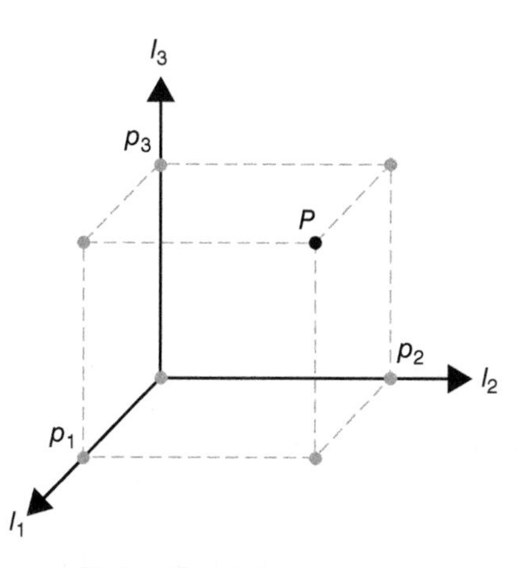

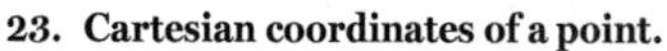
23. Cartesian coordinates of a point.

coordinates (0, 0, 0) consists of all points P such that $d(P, O) = r$ or, if the coordinates of P are (x, y, z),

$$x^2 + y^2 + z^2 = r^2.$$

The Witch of Agnesi depicted in Figure 22 has equation

$$(x^2 + b^2)y = b^3,$$

where $b = |OB|$ is the diameter of the circle used to define the Witch, and we have chosen the point O to be the origin and the point B to have coordinates $(0, b)$. The Witch is then the set of all points P with coordinates (x, y) which satisfy this equation.

Higher dimensions

The approach to geometry which replaces the ruler-and-compass constructions by equations is called *analytic geometry*. Note that, as the Witch lies on a plane, it is sufficient to use only two coordinates instead of three. The Cartesian coordinate system in this case is the x–y plane which you may recall from school. While the surface of the sphere is a shape embedded in three dimensions, a curve like the Witch of Agnesi is a shape embedded in two dimensions.

How about geometric shapes in higher dimensions? Most (but by no means all) professional mathematicians find visualizing higher dimensions as difficult as non-mathematicians do. Analytic geometry offers a remedy. Points on the plane can be identified with pairs of real numbers. We therefore denote the plane by $\mathbb{R}^2 = \mathbb{R} \times \mathbb{R}$, where $\mathbb{R}$ is the set of all real numbers and the symbol $\times$ is called the Cartesian product. Similarly, the Euclidean space $\mathbb{R} \times \mathbb{R} \times \mathbb{R}$ is denoted by $\mathbb{R}^3$. Its points correspond to triples of real numbers. Four-dimensional space $\mathbb{R}^4$ can therefore be *defined* as the set of quadruples of real numbers: a point in $\mathbb{R}^4$ is uniquely specified by its four coordinates (p_1, p_2, p_3, p_4). There is no need to imagine what the fourth dimension looks like, as long as one is willing to do the algebra. Somewhat less formally, each dimension corresponds to a degree of freedom which can be varied without altering the remaining dimensions. For any natural number n, we define points in n-dimensional Euclidean space as lists of n real numbers $(p_1, p_2, \dots, p_n)$. The adjective 'Euclidean' refers to the fact that an extension of the familiar Pythagorean theorem still holds and can be used to define the distance between two points P and Q in $\mathbb{R}^n$ as

$$d(P, Q) = \sqrt{(p_1 - q_1)^2 + (p_2 - q_2)^2 + \cdots + (p_n - q_n)^2}.$$

This analytic approach to higher dimensions enables us to answer seemingly impossible questions like:

> *What is the intersection of two spheres of unit radius in five dimensions whose centres are a unit distance apart?*

without a blink. First, extend the definition of the sphere of unit radius to $\mathbb{R}^5$: it is the set of all points with coordinates (x, y, z, u, w) such that

$$x^2 + y^2 + z^2 + u^2 + w^2 = 1,$$

where I have chosen the Cartesian coordinate system so that the centre of this sphere is at the origin. I shall now take the centre of the second sphere to be the point with coordinates $(0, 0, 0, 0, 1)$ in $\mathbb{R}^5$. In this way, the two centres are at a unit distance apart. This sphere is also required to have radius 1, so its equation is

$$x^2 + y^2 + z^2 + u^2 + (w - 1)^2 = 1,$$

where we have used the Pythagorean theorem to compute the squared distance between the centre and points on the sphere. To find the intersection of the two spheres, we need to determine all points with coordinates that satisfy both equations. This can be done by subtracting the second equation from the first. Simple algebra then gives $w = 1/2$. Substituting this value of w back into either of the equations yields

$$x^2 + y^2 + z^2 + u^2 = \frac{3}{4}.$$

This is an equation of a sphere of radius $\sqrt{3}/2$ in four-dimensional space. Thus a sphere in a four-dimensional space is the intersection of the two spheres in our problem. Easy? Given some practice with algebraic manipulations, it actually is quite straightforward.

A more interesting question is whether these manipulations still amount to geometry. There are purists among geometers who would prefer to stick to rulers, compasses, and pictorial intersections of shapes. A possible objection to the analytic methods could also be that the results seem to depend on various choices that we made along the way. Would we still have obtained a sphere in four dimensions as the answer to our question if we had positioned the centres of our two spheres at different points? This is a valid question, but it turns out that all possible choices lead to the same answer, the only difference being a possibly shifted centre for the resulting sphere. We shall return to this point in Chapter 6 when we discuss geometry from the point of view of groups of transformations.

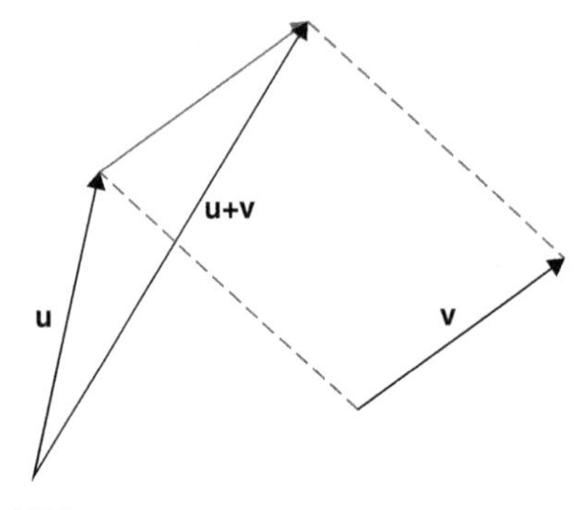

24. Vector addition.

Progress in geometry is often made using a combination of classical and analytic approaches. Pictorial representation leads to an intuition which may suggest what the answer could be. An explicit computation using Cartesian coordinates allows us to pin down the details and get further information, like the radius $\sqrt{3}/2$ of the sphere in our example.

Vectors

Vectors are a relatively recent concept going back to the 19th century, and vector methods are most effective when combined with the notation and formalism of algebra as employed in geometry.

A vector is an arrow attached to a point in space. If two arrows have the same length and point in the same direction, then they represent the same vector. I shall denote vectors by bold symbols, **u**, **v**, ...To add two vectors, **u** and **v**, take the arrow **v** and move it in space without changing its direction or length until its tail touches the head of the arrow **u**. Then join the tail of **u** to the head of **v** by an oriented line segment. The result of this operation is a vector denoted by **u**+**v** (Figure 24).

By instead moving the arrow **u** to the head of **v**, we can convince ourselves that for all vectors **u** and **v**:

V1 Addition is commutative: **u**+**v**=**v**+**u**.

Adding two vectors of the same length but with opposite direction gives the zero vector **0**, which has no direction, has zero length, and satisfies:

V2 There exists a zero vector **0** such that **u**+**0**=**u**.

The Cartesian coordinate system makes vector addition particularly simple. Let us say that a vector **u** is the arrow joining two points P and Q in the Euclidean space $\mathbb{R}^3$, with its tail at P and its head at Q. This vector is justifiably called $\overrightarrow{PQ}$. If the Cartesian coordinates of P and Q are (p_1, p_2, p_3) and (q_1, q_2, q_3), then the vector $\overrightarrow{PQ}$ can be represented by a triple of numbers called the vector components $(q_1 - p_1, q_2 - p_2, q_3 - p_3)$, which are the differences between the coordinates of Q and P. This leads to another perspective on vectors, this time in arbitrary dimension n, as n-tuples of numbers. The components of the vector **u**+**v** arise as the sum of the components of **u** and **v**, which makes property **V1** obvious.

We can not only add vectors to each other but also multiply them by numbers. This has an effect of changing the length of a vector and either keeping its direction for a positive multiple or reversing the direction for a negative multiple. For example, if **u**=$(1, -3, 2)$, then 3**u**=$(3, -9, 6)$ and -2**u**=$(-2, 6, -4)$. The branch of mathematics called linear algebra is devoted to the study of vectors as abstract objects which can be added to each other and multiplied by numbers in a way which satisfies a number of axioms, including **V1** and **V2**. In linear algebra we are not concerned with how vectors arise in space.

Vectors are especially useful in mechanics, where they represent quantities like velocity, acceleration, and force, which have a magnitude and a direction. Other quantities such as temperature or mass, which do not have a direction, are called scalars.

Adopting vectors in mechanics pays off when there are several forces acting on an object. Rather than analysing the effect of each force separately, we add the corresponding vectors to each other. The sum of these vectors is the net force $\mathbf{F}$. According to Newton's second law, $\mathbf{F} = ma$, the acceleration $\mathbf{a}$ resulting from this force is the vector obtained by multiplying $\mathbf{F}$ by a scalar $1/m$, where m is the mass of the object.

Hard problems with easy statements

Euclid's *Elements* has, until recently, been used for teaching geometry at high school and university levels. The text was written about 2,300 years ago, and while it is not free of fallacies (Euclid made some implicit assumptions about the ordering of points on line segments and about continuity, which were later added to the set of axioms) most of it is correct—these books are here to stay. Imagine teaching biology, chemistry, or physics from a text of similar age. In these fields, anything written more than 50 years ago is regarded as outdated, and often wrong. This is because the science is moving forward. And yet geometry is also moving forward—I will have more to say about this in the subsequent chapters—but once a correct proof of a theorem (such as the existence of five Platonic solids) has been presented, it becomes timeless. The theorem itself may generalize to other geometries or dimensions—indeed, such generalizations is what some professional mathematicians occupy themselves with—but none of this invalidates the original proof. Geometry is not unlike philosophy, where writings of the old masters are still studied, but this comparison may be unfair on both fields. You can either agree or disagree with the views of Immanuel Kant, and make a career out of this, as long as your arguments are presented in a coherent and eloquent way. Try, on the other hand, to disagree with the ancient Greeks and argue that $\sqrt{2}$ is rational after all! Perhaps art is a better analogy. Most would recognize the harmony of Leonardo da Vinci's work as timeless. There may be more to this comparison than meets the eye, and I shall return to it in Chapter 5.

While many theorems in geometry have been established by building on Euclid's axioms, there are still questions which are relatively easy to state, but less easy to settle. Many of these questions look like curious puzzles from a collection of recreational mathematics problems. It is remarkable that there is no straightforward way of deciding whether a geometric problem is just hard, but doable by a smart high school student with a clever idea, or one of the problems which professional mathematicians have tried to crack for decades or more. With this remark in mind, I invite readers to reflect on the following three problems:

Problem 1 Given any configuration of n points in the plane which do not all lie on one straight line, must there always be a straight line which contains only two points?

Problem 2 Can one paint a plane with three colours such that every point has some colour, and so that no two points with the same colour are a unit distance apart?

Problem 3 (Attempt once you have solved Problem 2.) What is the minimum number of colours needed to paint a plane so that no two points of the same colour are a unit distance apart?

Problem 1 was posed by James Joseph Sylvester—one of the most prominent English mathematicians of the 19th century. Sylvester could not solve it, and nor could anyone else until the mid 1930s. The following elementary proof was found in 1948 by Leroy Milton Kelly. Let $\mathcal{C}$ be the set of all points in Problem 1. Consider all pairs consisting of a point P in the set $\mathcal{C}$ and a line l that contains two other points in $\mathcal{C}$—call them Q and R—but misses P. The set $\mathcal{C}$ has finitely many elements, and so there are also finitely

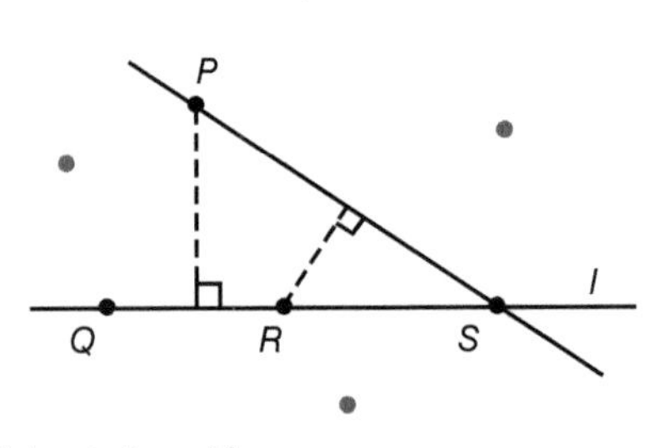

25. Solving Sylvester's problem.

many lines which connect points in $\mathcal{C}$. Therefore, there are finitely many such pairs (P, l), and we can pick one pair such that the distance from P to l is the smallest. Then the line l contains only two points. Otherwise, apart from Q and R, there would be at least one more point S on l and at least two out of the three points Q, R, and S must lie on the same side of the perpendicular joining the point P to the line l. Let us say that these two points are R and S, and that R is closer to the perpendicular than S (Figure 25). Then the distance from R to the line containing P and S is shorter than the distance from P to l. However, this is a contradiction, as we had assumed that the pair (P, l) minimizes all possible distances between points and lines. Therefore, no points other than Q and R can lie on the line l, and the Sylvester problem is answered affirmatively.

This proof, apart from solving Problem 1, demonstrates that elementary solutions may take a while to be found—an encouragement to amateur problem solvers, who are often frowned upon by mathematicians. In fact, there can be more than one line with the property that is mentioned in Sylvester's problem. It can be shown that, for any given number n, the number of such lines is at least $3n/7$. It is not known whether this bound can be improved.

The second problem is hard. While most people faced with it will sooner or later start drawing circles of unit radius, it is tricky to come up with a systematic way of doing it which shows that the answer to Problem 2 is a 'no'. And yet once the solution is revealed to them, they admit that it is very simple. Let us assume that it is possible to paint the plane with just three colours—call them red, blue, and white—in a way that is consistent with the requirements of Problem 2. Consider a point P on the plane—say it is white—and draw a circle C with unit radius centred at this point. All points on this circle have to be red or blue. Pick any two points on the circle which are a unit distance apart. One of them has to be red and the other blue, so we produced an equilateral triangle of unit side, and such that all the vertices of this triangle have different colours. Now reflect the white vertex in a line containing the blue and the red vertex. The resulting point must also be white. But this is true for all equilateral triangles constructed by picking two points on C which are a unit distance apart. Therefore, a circle centred at P and with radius $\sqrt{3}$ consists of points which are all white, and by taking a chord of unit length of this bigger circle we have constructed two white points which are a unit distance apart. We have now reached a contradiction, as originally we assumed that no such points can exist. This assumption must have been wrong—it is not possible to colour the plane with only three colours as in Problem 2.

The solution to Problem 3 is not known. The number of colours that the problem asks for is called the chromatic number of the plane. The solution of Problem 2 shows that this number is at least four. It is not greater than seven, which can be seen by considering the tiling of the plane with regular hexagons of side length d. Let us call the seven colours at our disposal (**1**, **2**, ..., **7**), and paint one of the hexagons with colour **1** and its six neighbouring hexagons with colours **2** to **7** in the clockwise direction, such that **7** is immediately on the left of **1**. Now translate this colouring to cover the whole plane (Figure 26).

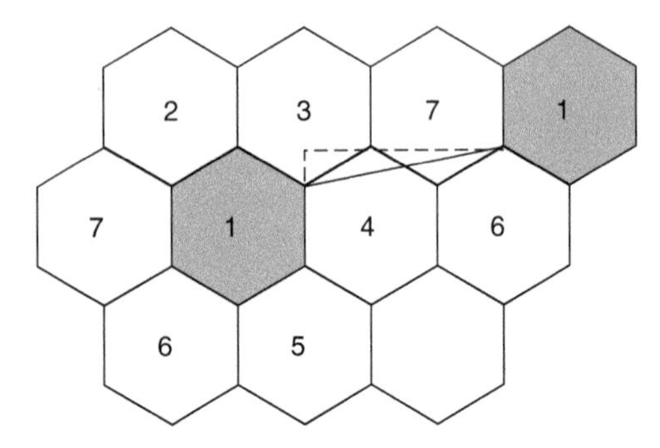

26. The chromatic number of the plane is at most 7.

Then an application of the Pythagorean theorem shows that no two points of colour **1** are at a unit distance apart if $\sqrt{7}/7 < d < 1/2$, so that picking $d = 3/7$ does the job. Therefore, the chromatic number of the plane is 4, 5, 6, or 7. This was the state of play until 2018, when a biologist and amateur mathematician, Aubrey de Grey, showed that this number cannot be four. He did it by a computer-assisted construction of a collection of 1581 points on the plane, which required five colours to avoid pairs of same-coloured points which are a unit distance apart. It is still not known what the chromatic number of a plane is.

Chapter 3
Non-Euclidean geometry

A sphere is the surface of a solid ball in three-dimensional Euclidean space. The distance between any pair of points on the sphere is the length of the shortest path connecting these two points and lying on the surface of the sphere. This last restriction is important—otherwise we would just connect the two points by a straight-line segment and compute its length using the Pythagorean theorem. The shortest paths turn out to be parts of great circles—the circles that arise as intersections of the sphere with planes through its centre (Figure 27).

Spherical geometry is the study of objects on the sphere. We define *lines* in this geometry to be great circles. These lines do not satisfy Euclid's parallel postulate in its form **E5'**: any two great circles intersect at exactly two antipodal points on the sphere. Thus spherical geometry is an example of a *non-Euclidean geometry*. It obeys Euclid's third and fourth postulates, but not the other three.

Another example of a non-Euclidean geometry we shall explore in this chapter is hyperbolic geometry. One of its models is the disc shown in Figure 3 in Chapter 1. With an appropriate reinterpretation of lines, this geometry also violates the parallel axiom while satisfying the other four. However, unlike the sphere, it cannot be embedded in ordinary space. Thus we shall introduce it as an abstract surface, then single out *lines* on this surface, and

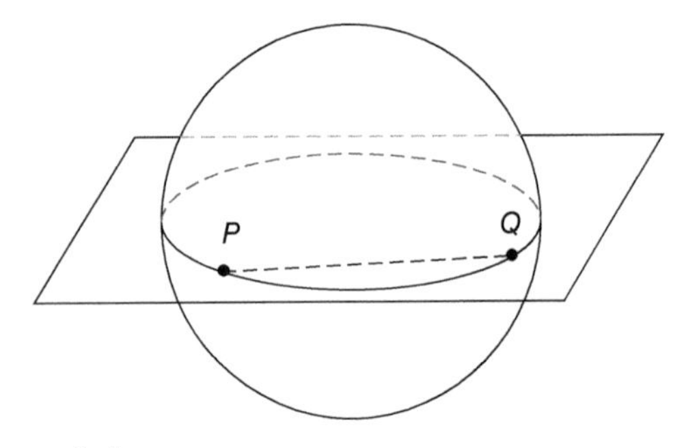

27. A great circle.

finally introduce the distance which makes these lines the shortest. This will allow us to reveal the apparent paradoxes of Escher's angels and devils in Figure 3.

Spherical geometry

The sum of the angles of any triangle on a plane is 180°, or π radians; our proof of this, given in Chapter 2, used the parallel postulate. This postulate does not hold in spherical geometry, so there is no reason to expect that spherical triangles have the same properties as Euclidean ones. In fact, the sum of angles in a spherical triangle is always greater than π. For example, if one of the vertices A is placed at the north pole and the other two vertices B and C are on the equator, such that the distance between them is a quarter of the length of the equator, then all three angles in the triangle ΔABC are $\pi/2$ radians or 90° (Figure 28). This is because the three segments of the triangle, AB, BC, and AC, are intersections of the sphere with three planes AOB, BOC, and AOC, where the point O is at the centre of the sphere and any two of the three planes intersect at a right angle.

The general formula for the sum of angles α, β, and γ of a triangle on a sphere with radius r is

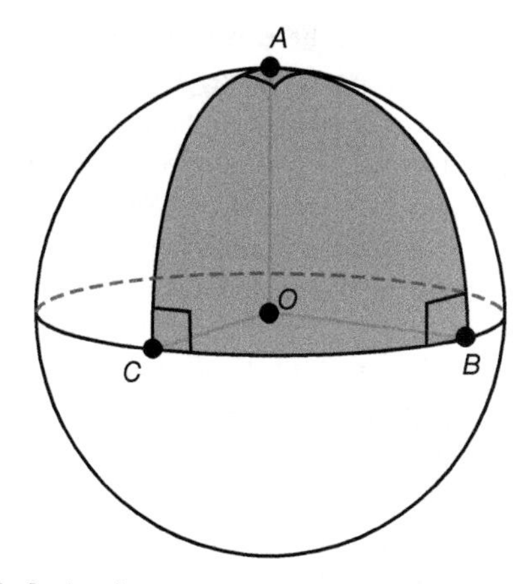

28. A spherical triangle.

S1
$$\alpha + \beta + \gamma = \pi + \frac{\text{Area}(\Delta ABC)}{r^2}.$$

In our example above, the area of the triangle ABC is one eighth of the surface area of the whole sphere, which gives $4\pi r^2/8 = \pi r^2/2$. Therefore, the r^2 cancels between the numerator and the denominator in **S1** and leaves the sum of angles to be $3\pi/2$ radians.

The formula **S1** has several striking consequences. For example, in spherical geometry there is no concept of similarity. In Euclidean geometry, two triangles are similar if they have the same angles but different sizes and therefore different areas. In spherical geometry the angles determine the area. To put it differently, the plane has no preferred distance scale—recall that we needed to add a choice of a distance ruler to Euclid's axioms, and different choices led to different units of length. On the other hand, the sphere has a natural absolute distance scale given by the radius of the sphere, or equivalently by the length of any of its equators divided by 2π. While the radius is an extrinsic concept, as it refers to the way the sphere sits in Euclidean three-dimensional space, the equatorial distance is intrinsic to the sphere. We shall return to this when we discuss Gaussian curvature in Chapter 4. Similarly, while there is no natural area scale in Euclidean geometry, as the area of the entire plane is infinite, a sphere does have an absolute scale—its own surface area which is equal to $4\pi r^2$. We can therefore measure an area of a spherical triangle or any other region of the sphere by comparing it with the total area. This underlies formula **S1**.

Another consequence of **S1** is the impossibility of making a plane map of any portion of the sphere which accurately represents all distances. In Chapter 2 such maps were called isometries. If an accurate a map were to exist, it would have to preserve distances as well as angles. Therefore, the segments *AB*, *BC*, and *CA* of great circles would be represented by straight-line segments joining the images of points A, B, and C on the map. The spherical triangle would map to an ordinary Euclidean triangle in a way which preserves the sum of angles, which is π. But this is impossible, as according to the formula **S1** the spherical triangle would then need to have zero area.

This non-existence of an isometry between S^2 and $\mathbb{R}^2$ has troubled map-makers since the earliest days of marine navigation. Modern-day cartography goes back to a map constructed by the Flemish geographer Gerardus Mercator in 1596. Mercator's map is made by placing a sphere inside a cylinder so that the equator is tangent to the cylinder. He then projected the parallels (intersections of the sphere with planes parallel to the equatorial plane) to the cylinder in a way which preserves the angles, and finally unravelled the cylinder onto a plane (Figure 29). As the angles are preserved, the curves of constant compass bearing measured relative to the north pole are represented by straight lines on a Mercator map. While these curves are usually not the shortest paths, they are easy to follow: a ship sails along a constant compass bearing from P to reach the destination Q. In the special case, when the points P and Q lie on the same meridian, the rhumb lines coincide with segments of

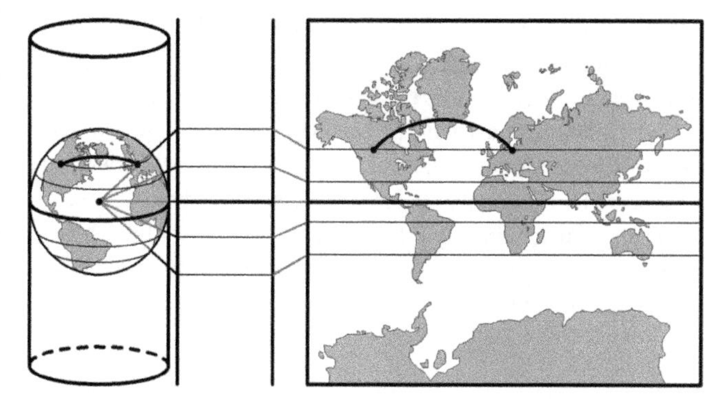

29. Mercator's projection.

great circles, and thus are indeed optimal. If you ever experienced a long-haul plane journey across the Atlantic in the days when the in-flight entertainment was a large screen displaying the flight trajectory, you will have noticed that the plane followed a curved path on the map (Figure 29). The map would have been a Mercator projection, and the flight path an image of a spherical line on this map.

Regions far away from the equator are distorted on Mercator's map—Greenland appears to be the same size as Africa, whereas Africa is about 15 times bigger. While the Mercator map preserves angles, it does not preserve areas. There are other types of maps which preserve areas but distort distances and great circles to the extent that they are useless for navigation. One such map is obtained by placing the sphere inside a cylinder and projecting it outwards from the axis of the cylinder. Archimedes proved that the area of any region of the sphere is the same as the area of the projection of this region onto the cylinder, and therefore it is the same on the map obtained by cutting the cylinder along one of its parallel lines. This result was engraved on Archimedes' tombstone when he died in 212 BC.

The formula **S1** implies that something has to give, as a perfect map cannot be constructed. In fact, even if one did not care about distances, angles, and areas being represented accurately, it would not be possible to construct a projection from the sphere to a single map which is continuous and such that all points on the sphere are represented on the map and, conversely, each point on the map corresponds to a unique point on the sphere. To cover the whole sphere, an atlas which consists of at least two maps is needed. We return to this point in Chapter 4, when we discuss manifolds.

What remains is to prove the formula **S1**. Once the radius r of the sphere is fixed, we can choose a distance ruler in $\mathbb{R}^3$ such that r is the unit distance. We can therefore set r to 1. Having made this choice, the angular units of measure are related to the absolute distance and area scales on the sphere by

$$\pi \text{ radians} = \frac{1}{2} \text{ (equatorial length)} = \frac{1}{4} \text{ Area}(S^2).$$

Step 1 Consider a region of the sphere bounded by two great circles connecting two antipodal points, A and A'. Such a region is called a lune (Figure 30(a)).

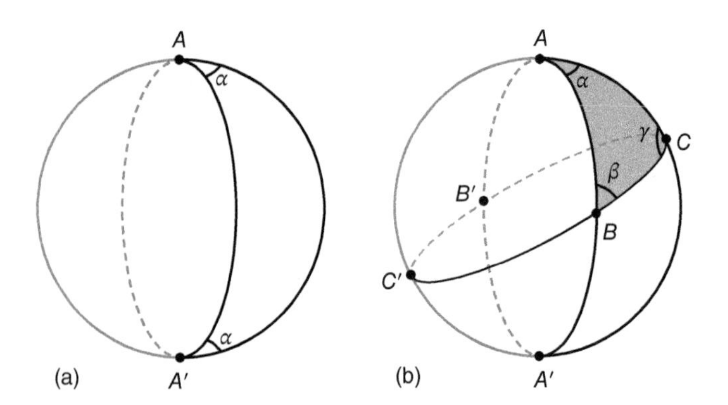

30. A lune.

If the angle between the great circles is α, then the area of the lune is

$$\text{Area}(S^2) \times \frac{\alpha}{2\pi} = 2\alpha,$$

assuming that the sphere has unit radius, so that its area is 4π.

Step 2 The lune from Step 1 is a union of two spherical triangles (Figure 30(b)), ΔABC and $\Delta A'BC$, so that

$$\text{Area}(\Delta ABC) + \text{Area}(\Delta A'BC) = 2\alpha.$$

Proceeding analogously with the segments connecting antipodal pairs (B, B') and (C, C') gives

$$\text{Area}(\Delta ABC) + \text{Area}(\Delta AB'C) = 2\beta,$$

$$\text{Area}(\Delta ABC) + \text{Area}(\Delta ABC') = 2\gamma.$$

Adding these three equations yields

$$3\text{Area}(\Delta ABC) + \text{Area}(\Delta A'BC) + \text{Area}(\Delta AB'C) + \text{Area}(\Delta ABC') = 2(\alpha + \beta + \gamma).$$

Step 3 Triangles ΔABC, $\Delta ACB'$, $\Delta AB'C'$, and $\Delta AC'B$ make up a hemisphere bounded by the equator containing the points B, C, B', and C'. The area of this hemisphere is 2π. The triangles $\Delta AB'C'$ and $\Delta A'BC$ are inversions of each other with respect to the centre of the sphere, and have the same areas. Therefore,

$$\text{Area}(\Delta ABC) + \text{Area}(\Delta A'BC) + \text{Area}(\Delta AB'C) + \text{Area}(\Delta ABC') = 2\pi.$$

Subtracting this equation from the last equation in Step 2 gives

$$\alpha + \beta + \gamma = \pi + \text{Area}(\Delta ABC),$$

which is the spherical sum of angles formula **S1** for a sphere with unit radius.

Hyperbolic geometry

Historically, the first example of a non-Euclidean geometry in which the parallel postulate is false and the Pythagorean theorem does not hold was hyperbolic geometry. It can be introduced in several equivalent (but not *obviously* equivalent) ways, ranging from listing a set of axioms to exhibiting a concrete model. I shall choose the latter, and present the model first put forward by the Italian Eugenio Beltrami in 1868 and subsequently rediscovered and popularized by the French mathematician Henri Poincaré, whose name it now carries. The arena for hyperbolic geometry in this approach is the unit disc: the interior of a circle with radius one. We assert that the following paths are hyperbolic lines (analogues of straight lines in Euclidean geometry, or great circles in spherical geometry):

1. Diameters of the boundary circle
2. Circular paths which intersect the boundary circle at 90°.

Any two points inside the disc can be connected by a unique hyperbolic line (Figure 31), which can be seen as follows.

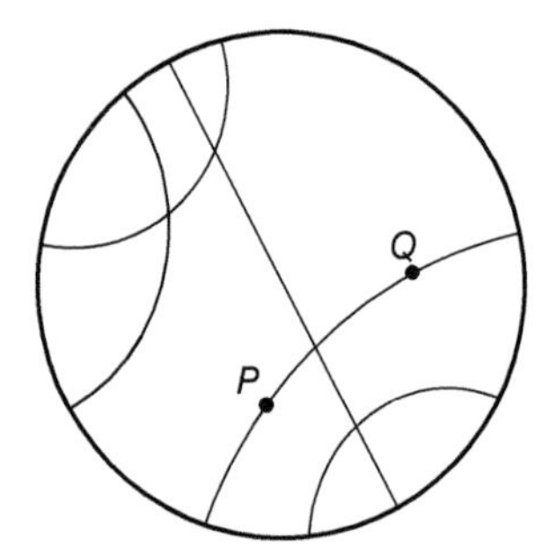

31. Hyperbolic lines.

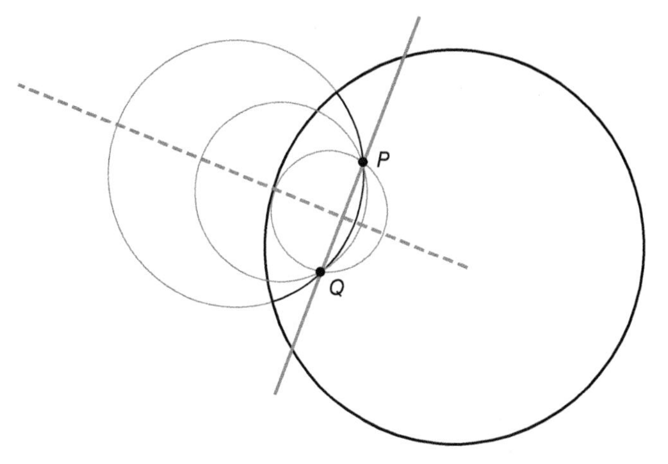

32. A construction of a hyperbolic line.

If P and Q lie on the diameter of the boundary circle, then this diameter is the hyperbolic line. Otherwise, consider all circles containing P and Q, whose centres lie on the perpendicular bisector of PQ. One of these circles, call it $\mathcal{C}$, is tangent to the boundary circle, so the angle between the two is 0°. Now slide the centre of $\mathcal{C}$ away from the centre of the boundary circle, making sure that $\mathcal{C}$ contains P and Q. Each of the resulting circles intersects the boundary circle at two points at an angle which varies between 0° and the angle that the Euclidean secant containing PQ makes with the boundary circle. The latter is greater than 90°, as we have assumed that P and Q do not lie on the diameter. Thus, by continuity, there is exactly one circle containing P and Q and intersecting the boundary at 90° (Figure 32).

Given a hyperbolic line l and a point P inside the disc and not on l, there are infinitely hyperbolic lines through P which do not intersect l. These hyperbolic lines are called parallel to l (Figure 33).

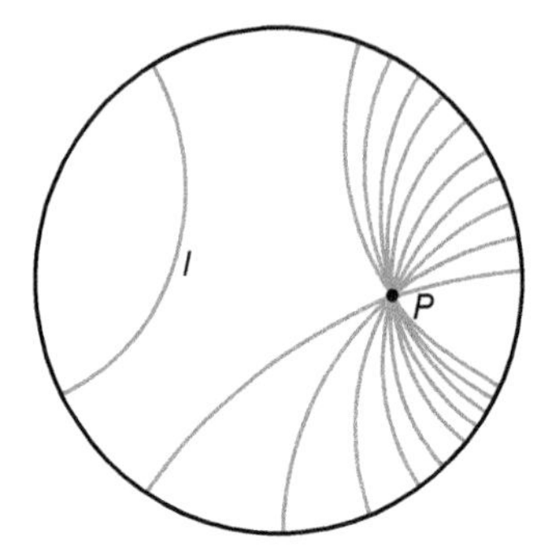

33. Infinitely many parallel lines.

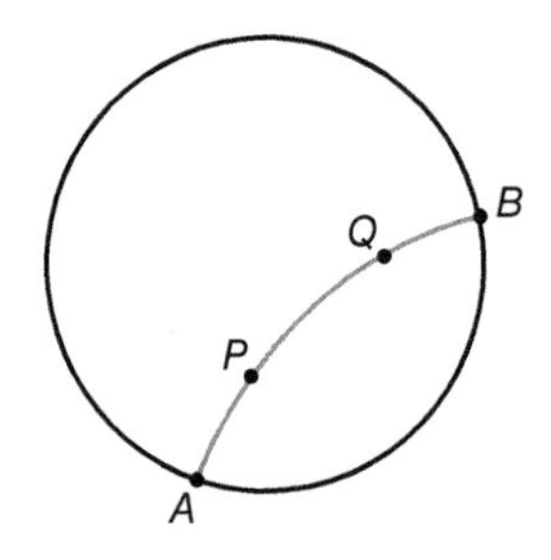

34. The hyperbolic distance.

In Euclidean geometry there is only one such line (this is Euclid's postulate **E5'**), and in spherical geometry there are none, as any two great circles intersect at exactly two points.

Having defined the hyperbolic lines, we ask whether there is a notion of distance on the Poincaré disc which makes these lines the shortest paths. The answer to this is 'yes', and the expression for the hyperbolic distance between two points P and Q is:

H1 The distance formula (Figure 34),

$$d(P, Q) = \ln\left(\frac{|AQ||BP|}{|AP||BQ|}\right),$$

where A and B are points at which the hyperbolic line through P and Q meets the bounding circle. In the formula **H1**, the expressions $|AP|$, etc. denote the Euclidean distance between A and P, and ln is the natural logarithm to base e, where

$$e = 1 + \frac{1}{1} + \frac{1}{1\cdot 2} + \frac{1}{1\cdot 2\cdot 3} + \cdots = 2.7182\ldots,$$

so $a = \ln b$ if and only if $b = e^a$ for any two real numbers a and b such that $b > 0$.

To get a better sense of the formula **H1** and develop some intuition of hyperbolic geometry, let us compute the hyperbolic distance when one of the two points is the origin, and so the hyperbolic line is part of a diameter of the bounding circle. We therefore choose P to have coordinates $(0, 0)$ and Q to have coordinates $(r, 0)$, where r is some number between 0 and 1 (I have aligned the coordinates so that the horizontal axis contains Q—there is no loss of generality in doing so, as the rotations of the disc around the origin preserve the hyperbolic distance). The diameter of which PQ is a part intersects the bounding circle at two points: A with coordinates $(-1, 0)$, and B with coordinates $(1, 0)$. Therefore,

$$|AQ| = 1 + r, \quad |BP| = 1, \quad |AP| = 1, \quad \text{and} \quad |BQ| = 1 - r,$$

and $d(P, Q) = \ln\left(\frac{1+r}{1-r}\right)$. This is a function of r, which we can plot on a graph (Figure 35).

If the point Q is close to the disc boundary, then r is close to 1 and the denominator in the logarithm becomes small, resulting in a large hyperbolic distance. In the limiting case, when $r = 1$ the distance is infinite. Thus in hyperbolic geometry the boundary circle is infinitely far away from the centre of the disc. Note that the Euclidean distance between the two is equal to 1. To get even more insight into what is going on, let us consider another point R on the diameter lying between P and Q such that the Euclidean distance between Q and R is δ and the coordinates of R are $(r - \delta, 0)$. You may want to repeat my calculation of $d(P, Q)$ to find that the closer Q is to the boundary (or equivalently the closer r is

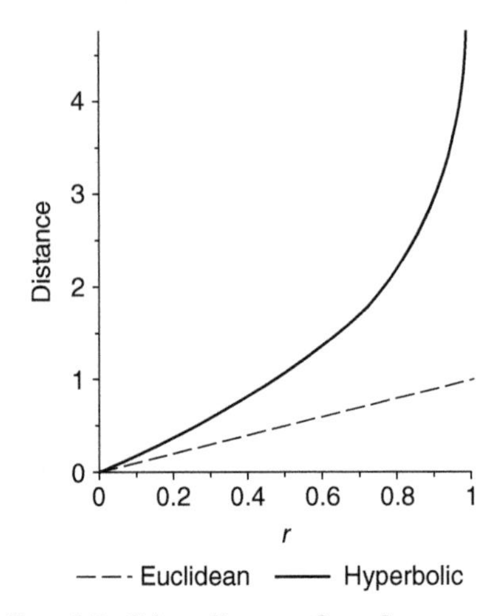

35. Hyperbolic and Euclidean distances from the centre of the disc.

to 1), the larger the hyperbolic distance, despite the fact that the Euclidean distance between the two points is always the same, and equal to δ. This explains the apparent paradox in M. C. Escher's *Circle Limit IV* illustrating the hyperbolic disc (Figure 36): the devils close to the boundary circle in the Figure appear smaller than those near the centre of the disc, but the height of each devil from head to tail is the same when hyperbolic distance, rather than Euclidean distance, is used.

The hyperbolic area is also the same for all devils. There are infinitely many devils inside the disc, so the hyperbolic area of the disc is infinite. Any devil is a finite (though a possibly large) distance from the centre of the disc, and we have shown that the centre is infinitely far away from the boundary. Therefore, the triangle inequality,

$$d(\text{centre, devil}) + d(\text{devil, boundary}) \geq d(\text{centre, boundary}) = \infty,$$

implies that all devils, even those which appear close to the boundary from the Euclidean perspective, are actually infinitely

36. ***Circle Limit IV*** **(M. C. Escher).**

far from the boundary circle. It may seem to us from the Euclidean perspective that the devils reside inside a circle, but for them there is no 'outside', as the circle is infinitely large from a hyperbolic perspective.

Hyperbolic polygons

We now turn to triangles in hyperbolic geometry and check whether or not their angles add up to π (Figure 37). The triangle in Figure 37 consists of three hyperbolic line segments connecting vertices A, B, and C. The angles of this triangle add up to less than π, which is apparent if we consider the Euclidean triangle with the same vertices. The two hyperbolic lines emanating from each vertex are inside the wedge formed by the two Euclidean line segments, so all hyperbolic angles are smaller than the corresponding Euclidean angles.

The sum of the angles in a hyperbolic triangle is always smaller than π. Deducing the precise formula for the sum of the angles from the distance function **H1** is somewhat complicated, and the formula was first 'guessed' by the 18th-century Swiss

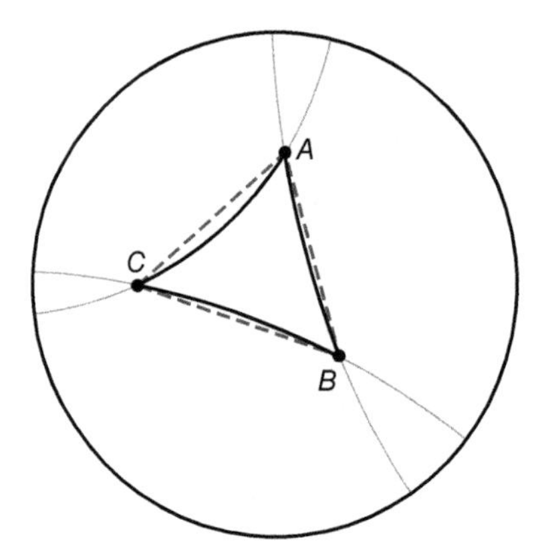

37. A hyperbolic triangle.

mathematician Johann Lambert. While we cannot be sure of the exact steps of his reasoning, Lambert regarded the hyperbolic geometry as spherical geometry on a sphere with imaginary radius. Substituting $r = i$, where $i^2 = -1$, in the spherical angle formula **S1** gives

H2 $$\alpha + \beta + \gamma = \pi - \text{Area}(\Delta ABC).$$

A hyperbolic triangle can have all three angles equal to zero. This happens only if all the vertices of the triangle lie on the boundary circle, as then any pair of hyperbolic lines emanating from a given vertex must be tangent to each other, since the lines intersect the boundary at 90°. Triangles with all zero angles are called *ideal*, and the formula **H2** implies that the area of an ideal triangle is π. This is the largest possible area of a triangle in hyperbolic geometry, while in Euclidean geometry there is no upper limit for an area. A hyperbolic disc can also be tiled with infinitely many ideal triangles (Figure 38).

In Chapter 2 we saw that we can tile the Euclidean plane with regular polygons in only three different ways: with triangles, with

38. Tiling by ideal triangles.

six of them meeting at each point, with squares with four of them meeting at each point, and with hexagons, with three of them meeting at each point. Thus, if k regular n-gons meet at a point, then the possible pairs (n, k) are $(3, 6), (4, 4)$, and $(6, 3)$, and in each case

$$\frac{1}{n} + \frac{1}{k} = \frac{1}{2}.$$

The theory of hyperbolic tilings by regular n-gons is much richer, and in fact any n-gons can be used as long as

H3
$$\frac{1}{n} + \frac{1}{k} < \frac{1}{2}.$$

To see how this comes about, consider k regular n-gons meeting at a point, so that the internal angle is $2\pi/k$. Another way to compute this internal angle is to partition the n-gon into $(n-2)$ triangles—in Euclidean geometry this is shown in Figure 16. Comparing the two resulting expressions for the internal angle, and using **H2**, we have

$$n\frac{2\pi}{k} = (n-2)(\pi - \text{Area}),$$

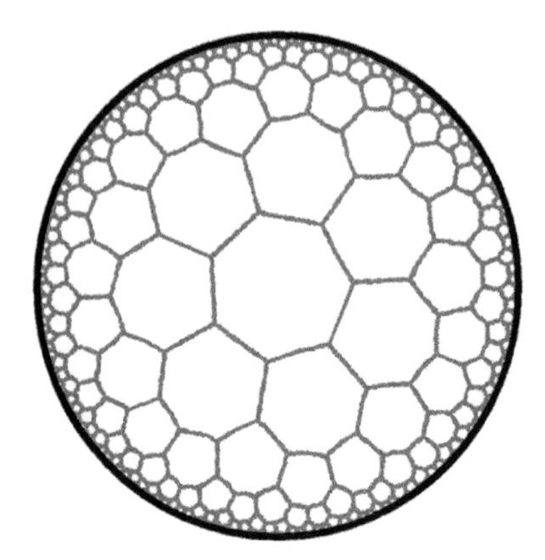

39. Tiling with regular heptagons.

where Area is the total area of the n-gon, which is positive. Dividing both sides of this expression by $2\pi n$ and rearranging gives **H3**, and so **H3** is necessary for a hyperbolic tiling of type (n, k) to exist. It is also sufficient, as given a pair (n, k) satisfying **H3**, the area of the n-gons can be chosen so that k of these n-gons meet at a point, leaving no gaps. We may assume that this point is the origin of the disc, and then fill in the rest of the disc by reflecting each of the existing k n-gons with respect to one of its outer sides: a reflection maps a point P to a point P' such that P and P' lie on a hyperbolic line which intersects the line of reflection at the right angle, and the line of reflection divides the segment PP' in two. Iterating this process covers the whole disc with n-gons of the same area, and without gaps. A disc can be tiled with regular heptagons (Figure 39). The heptagons in this figure are indeed regular: the hyperbolic lengths of all sides are equal.

Euclid's parallel postulate

Euclid's parallel postulate was a subject of controversy for over 2,000 years. Many mathematicians tried to deduce the parallel postulate from the other four axioms, but all attempts failed. In the early 19th century it was realized that one should instead agree that Euclid was, after all, right in including this postulate as an independent axiom of his geometry. There can exist a geometry different from the Euclidean one, where the first four axioms hold but the fifth does not. The first to arrive at this conclusion was possibly Gauss, who coined the name *non-Euclidean geometry*. Apparently wary of the controversy that hyperbolic geometry may cause, Gauss never announced his results, and the first papers on the subject were published by the Russian Nikolai Lobachevsky in 1829 and the Hungarian János Bolyai in 1832. Concrete models of hyperbolic geometry, including the disc model discussed in this chapter, were given by Beltrami in 1868.

In spherical geometry it is not only the parallel axiom that fails, but also two other axioms. Hyperbolic geometry differs from Euclidean geometry only in that the parallel axiom does not hold. Thus the existence of hyperbolic geometry proves that it is impossible to deduce the parallel axiom from the first four, for if it could be deduced, then it would also hold in hyperbolic geometry, which is not true. Some theorems in Euclidean geometry also hold in hyperbolic geometry; such theorems can be established without referring to the parallel axiom. For example, in both geometries, every exterior angle of a triangle is greater than each of the opposite interior angles. This theorem is not true in spherical geometry.

I have introduced the Poincaré disc in a strange way: not as a surface in Euclidean three-dimensional space which is then mapped to some region of a plane (that is how we would think about the sphere—the fact that the segments of great circles are the shortest paths then follows from the Pythagorean theorem in the three-dimensional Euclidean space), but rather as an abstract model consisting of a preferred set of lines and a distance function which makes these lines the shortest. The Poincaré disc, with its hyperbolic lines, satisfies the first four axioms of Euclid, but not the fifth one. Does this disc really exist? You might rightly object at this stage, for if the axioms are satisfied and no contradiction

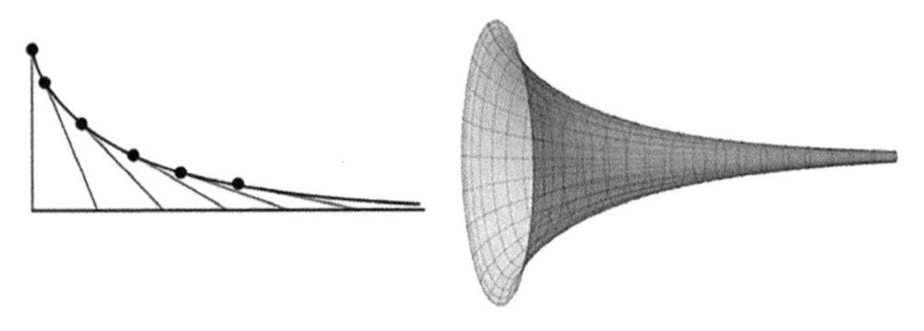

40. The tractrix and the pseudosphere.

follows from the model, then it does exist in the abstract Platonic world of mathematics. But what I am asking is whether we could have arrived at a hyperbolic disc in a way that is similar to how one defines a sphere: as a surface in Euclidean three-dimensional space with some special property. The answer to that is 'no'. There exist surfaces in $\mathbb{R}^3$—the so-called pseudospheres—which have properties similar but not identical to that of the hyperbolic disc; for example, the angle formula **H1** holds for all triangles. One construction of a pseudosphere goes as follows (Figure 40):

Step 1 Consider a vertically oriented rod of length r on the Euclidean plane, with a weight attached to the rod at one end. Pull the other end along the horizontal axis. The curve traced by the weight is called a tractrix.

Step 2 Revolve the tractrix about the horizontal axis. The resulting surface is a pseudosphere (Figure 40).

The area of the pseudosphere is finite, unlike that of the Poincaré disc, despite its having the infinitely long trumpet. Its surface area is equal to $4\pi r^2$ which is the same as for the sphere with radius r, and justifies the terminology. The triangles constructed from the shortest line segments on the pseudosphere satisfy **H1**, but there is a difference between the pseudosphere and the disc: on the disc one can follow a hyperbolic line indefinitely, without ever getting to the boundary, while on the pseudosphere the hyperbolic lines end abruptly on the boundary circle. So the pseudosphere only gives a local model of hyperbolic geometry.

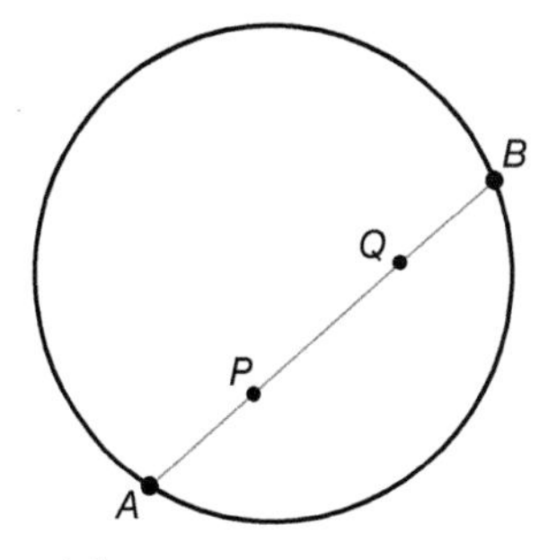

41. The Beltrami model.

There are other models of this geometry which are global. The following model was discovered by Beltrami in the same paper as the Poincaré disc was introduced. As for the Poincaré disc, it consists of a disc—call it $\mathbb{B}$—bounded by a circle of unit radius. The hyperbolic lines in $\mathbb{B}$ are defined differently, as Euclidean lines cutting the boundary at two points (Figure 41). It is clear from this definition that any two points in $\mathbb{B}$ are connected by a unique line. The angles in the Beltrami model do not correspond to the Euclidean angles, and the distance in $\mathbb{B}$ is equal to

$$d(P,Q) = \frac{1}{2}\ln\left(\frac{|AQ||BP|}{|AP||BQ|}\right),$$

which is half of the expression **H1** for the distance in the Poincaré disc. Beltrami has shown that the two models are equivalent.

The Poincaré disc, the Beltrami model, and the pseudosphere all differ from Euclidean geometry by the presence of curvature. This curvature is the same, and in fact negative, for all three models of hyperbolic geometry. In the next chapter we discuss the concept of curvature for general surfaces.

Chapter 4
Geometry of curved spaces

Imagine a civilization of intelligent bugs living on a plane curve. These bugs have developed sophisticated mathematical techniques which go beyond the scope of this book. They also have access to precise measuring apparatus. Can they determine whether their one-dimensional universe is in fact a curve, rather than a straight line (Figure 42)? It turns out that they cannot. The curvature of a curve is an extrinsic property. It is visible to us, because we view the curve as it sits in the plane. We, can however, delicately straighten the curve without changing its overall length or altering distances between the points. The bugs would just not notice!

Now imagine another advanced civilization of bugs, but this time living on the surface of a sphere. It turns out that these bugs can, just by walking around and never leaving the surface, decide that their universe is curved. Curvature is an intrinsic property of the surface. This is far from obvious. The German mathematician Carl Friedrich Gauss, who is widely regarded as one of the greatest mathematicians of all time, was very proud of this discovery, which he made in 1828. He called it the *Theorema Egregium*, which translates from Latin as 'Remarkable Theorem'. In this chapter we shall, starting from the extrinsic curvature of curves, define Gaussian curvature and explain how it can be computed intrinsically. We shall also look at higher-dimensional

42. Extrinsic curvature of a curve.

generalizations of surfaces called *manifolds*, and introduce *Riemannian geometry*.

Curved curves

We shall define the curvature of a curve on the Euclidean plane. Let this curve be a circle of radius r. In the absence of a formal definition, you will agree that the circle is curved in the same way at any of its points. Moreover, the larger the radius r, the less curved the circle. Circles of very large radius resemble straight lines, for which we would expect the curvature to vanish. We shall therefore define the curvature of a circle to be the reciprocal of its radius,

$$\kappa = \frac{1}{r}.$$

In the limit where $r \to \infty$, we have $\kappa \to 0$, which is the curvature of a straight line.

How about curves other than circles (Figure 43)?

Let us call the curve on Figure 43 γ. The idea is to approximate γ by circles with radii varying from point to point. For any point P on γ, there are infinitely many circles which are tangent to γ at P. We shall select one of these—called the osculating circle—which gives the best fit, and define the curvature of γ at the point P to be the reciprocal of the radius of the osculating circle.

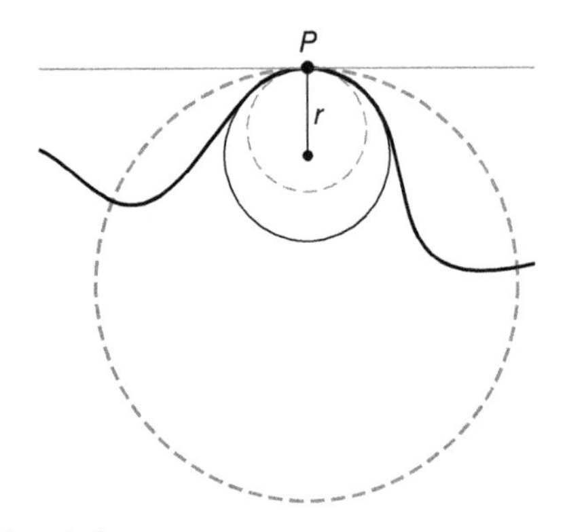

43. Osculating circle.

To identify the osculating circle, note that near the point P the curve divides the plane into two parts. Some of the circles tangent to γ at P lie entirely on one side, and some lie entirely on the other. The osculating circle separates these two kinds of circles. On Figure 43 it is drawn with a solid line.

If you are familiar with calculus, you will reconcile this pictorial description with the fact that, at the point P, the curve γ has the same first and second derivatives as the osculating circle. This defines the osculating circle and its radius uniquely. If the curve is given as a graph of a function $y = f(x)$, then its curvature is

$$\kappa = \frac{f''}{(1+(f')^2)^{3/2}},$$

where f' and f'' are the first and second derivatives of f with respect to x. For example, the curvature of the parabola $y = x^2$ (Figure 44) tends to zero, as x tends to infinity. This is in agreement with our intuition: both tails of the parabola look approximately flat. The curvature is biggest at the origin $(x = 0, y = 0)$, where it equals 2.

The approach to geometry based on calculus is called differential geometry. While calculus is a powerful technique and is required

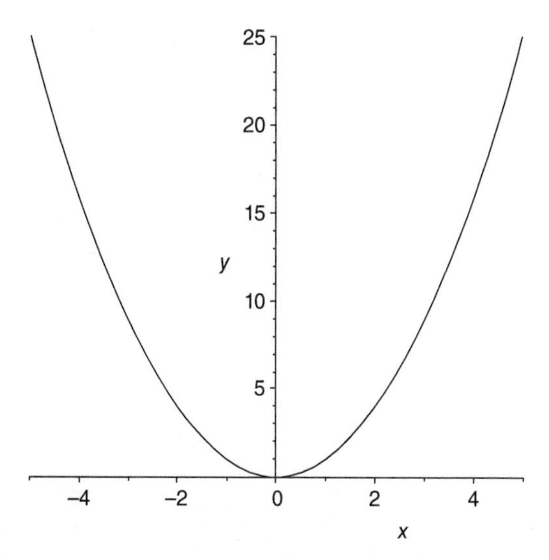

44. Curvature of the parabola $y = x^2$ is $\kappa = 2(1 + 4x^2)^{-3/2}$.

to prove most assertions which will be made in this chapter, I have—in order to keep the discussion at an elementary level—chosen not to use calculus, and to provide pictorial explanations instead. Readers familiar with calculus should consult books listed in the bibliography.

Curved surfaces

We shall use our definition of the curvature of a curve to define the curvature of a surface at any of its points. This surface, which I shall call Σ, is assumed to be smooth with no edges (Figure 45).

Step 1 Slice the surface by a plane perpendicular to it at some point P. Using the osculating circle, find the curvature of the slice curve which is the intersection of Σ with the plane.

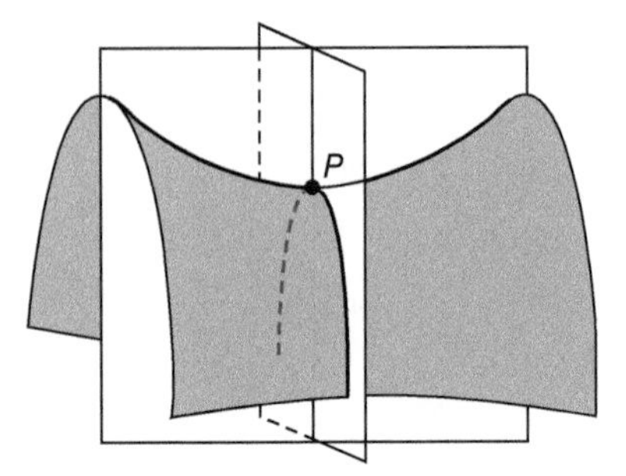

45. Slicing a surface.

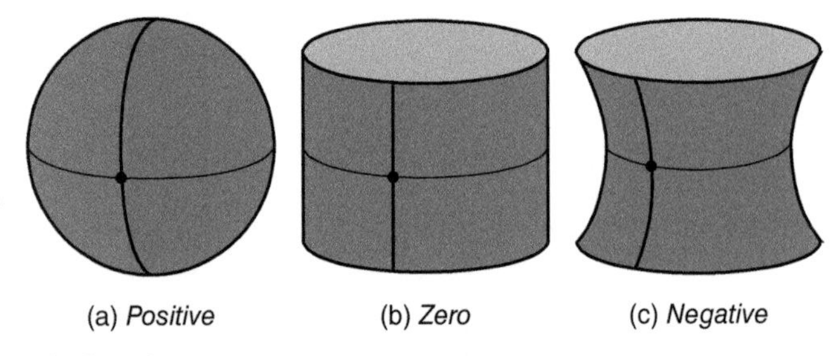

46. Gaussian curvature.

Step 2 Repeat the process by rotating the sliced plane, while keeping it perpendicular to Σ at *P*. This gives the curvatures of the slice curves in all directions. We take these curvatures with a positive sign if the curve slices downwards, away from the plane tangent at *P*, and with a negative sign otherwise.

Step 3 Let $\kappa_{\max}$ be the largest of the curvatures obtained in Step 2, and let $\kappa_{\min}$ be the smallest. Define the Gaussian curvature of the surface to be

$$K = \kappa_{\min} \times \kappa_{\max}.$$

The Gaussian curvature of the sphere with unit radius is the same at any of its points, and equal to 1 (Figure 46). All perpendicular planes intersect the sphere in circles of the same radius, so $\kappa_{\text{min}} = \kappa_{\text{max}} = 1$ and $K = 1 \times 1 = 1$. The cylinder in Figure 46(b) has zero Gaussian curvature, and is therefore called flat. The plane parallel to the axis intersects the cylinder in a straight line, and $\kappa_{\text{min}} = 0$ for all points on the cylinder. Therefore, $K = 0 \times 1 = 0$. The third surface in Figure 46(c) has negative Gaussian curvature. The perpendicular planes shown in Figure 46(c) intersect the surface at circles bending in opposite directions, so K is a product of a negative and a positive number.

Points on Σ at which $K > 0$ are called elliptic. If P is an elliptic point, then the tangent plane to Σ at P does not intersect the surface, at least in the small neighbourhood of the surface surrounding P. Points where $K < 0$ are called hyperbolic. The surface at such points resembles a horse saddle, or a mountain pass, where the terrain rises in some directions and drops in others. As a hiker walks from a hyperbolic point Q on the pass to an elliptic point P on the top of a mountain, the Gaussian curvature changes along the path from its negative value at Q to a positive value at P. If the path is sufficiently smooth, there will exist a point along it where the curvature is exactly zero. Near such a point, Σ looks like the surface of a cylinder.

To make sense of Gaussian curvature, we used the extrinsic properties of the surface, and looked at the way it sits embedded in the Euclidean space $\mathbb{R}^3$. Gauss's Theorema Egregium states that K can be defined in a way intrinsic to the surface, and can be measured at any point of Σ by a bug unaware of the surrounding three-dimensional space. One way this bug could do it is by picking a point, P, and constructing a circle with radius r about P. The notion of the distance d measured by the bug is intrinsic to the surface. For example, it would use the formula **H1** if Σ were the Poincaré disc. For a general surface, the circle $\mathcal{C}$ centred at P is defined to be

$$\mathcal{C} = \{\text{all points } Q \text{ on } \Sigma \text{ such that } d(P, Q) = r\}.$$

The circumference of this circle would be $2\pi r$ if Σ were the Euclidean plane. The presence of positive Gaussian curvature makes this circumference smaller and negative Gaussian curvature makes it larger, according to the formula

$$\text{circumference}(\mathcal{C}) = 2\pi r - K\pi\frac{r^3}{3} + \dots,$$

where the dots denote terms containing powers of r greater than three which can be neglected for small circles.

A typical surface, like that of a pear, is curved irregularly and its Gaussian curvature varies from point to point. The plane, the sphere, and the Poincaré disc are special, as their Gaussian curvatures are the same at every point. On these three surfaces, any triangle satisfies

$$\alpha + \beta + \gamma = \pi + K \times \text{Area}(\Delta ABC).$$

Euclid's third and fourth axioms are satisfied if $K \geq 0$, and the first four axioms hold if $K \leq 0$. The flat geometry of the plane, where Euclid's fifth axiom holds, arises as the limiting case of both hyperbolic and spherical geometry when $K = 0$.

The table below summarizes the differences between the hyperbolic, spherical, and Euclidean geometries. To make these differences more transparent, I shall make the following definition. For any pair (P, l), where l is a line and P is a point not on l, let us define $N(P, l)$ to be the number of lines through P which do not intersect l. In Euclidean geometry this number is equal to 1; this is one of the statements of Euclid's parallel axiom.

Geometry	Sum of angles in a triangle	Gaussian curvature	Total area	$N(P, l)$
Hyperbolic	$< \pi$	negative	∞	∞
Euclidean	$= \pi$	zero	∞	1
Spherical	$> \pi$	positive	finite	0

The Gauss–Bonnet theorem

Gaussian curvature is a local quantity which can be computed at any point P of Σ using small circles, and it does not depend on how the surface looks away from P. There is, however, an amazing connection between Gaussian curvature and the global properties of the surface. To explain this, I shall first give another intrinsic description of K. Looking back at the spherical and hyperbolic geometry formulae **S1** and **H2**, we see that, at least in these two cases,

$$K = \frac{\alpha + \beta + \gamma - \pi}{\text{Area}(\Delta ABC)},$$

where ΔABC is any triangle.

A way to make sense of this formula for a surface where K changes from point to point is to consider $\Delta(P)$ to be a small triangle centred at P. Such a triangle is almost Euclidean, so its centre can be defined as in Euclidean geometry. The Gaussian curvature at P then arises as a limiting case of the formula above, where the area of $\Delta(P)$ tends to zero and the sum of angles tends to π. Let us assume that Σ is a surface with no edge, which can be completely enclosed by a sphere with arbitrarily large radius. Spheres and doughnut-shaped surfaces belong to this category, but a disc does not, as it is bounded by a circular edge. Imagine covering Σ with small, but not necessarily identical, triangles in an arbitrary way. Each of these triangles can be used to compute K at their centre. Let us do that, and multiply the resulting K by the area of the corresponding triangle. Now sum all these numbers. There are infinitely many of them, and they are infinitely small (because the areas are) so one needs to employ calculus and replace the sum by an integral. The Gauss–Bonnet theorem states that the integral of K over the whole surface when divided by 2π is related to the number of holes g, also called the genus, of the surface by the formula

$$\frac{1}{2\pi}\int_{\Sigma} K = 2 - 2g.$$

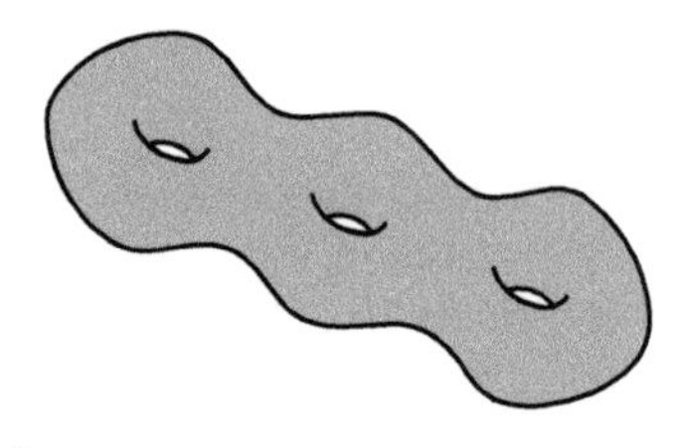

47. Genus 3 surface.

If, for example, Σ is a sphere with radius 1, then $K = 1$ and the integral gives the area of S^2 equal to 4π. Dividing this by 2π gives 2, which agrees with the statement of the theorem, as the sphere has no holes and its genus g equals zero. Not too surprising so far. But now imagine stretching and bending the sphere to form a pear-shaped surface. There will be regions where K is positive and regions where it is negative, but the total integrated curvature will balance out to give 4π. If Σ is a multi-holed doughnut (Figure 47), then K is positive close to the outside region and negative near the holes. The total curvature integrates to -8π, regardless of how the doughnut is bent.

Although rooted in 19th-century mathematics, the Gauss–Bonnet theorem is regarded as one of the deepest results in geometry, which has shaped a lot of the 20th- and 21st-century development in the subject. It connects local and global properties of surfaces, and has been generalized to Riemannian manifolds—higher dimensional analogues of surfaces. We shall discuss those next.

Manifolds

A manifold is a space which, in a small neighbourhood of any of its points, looks like the Euclidean space. The dimension of this Euclidean space—the number of Cartesian coordinates needed to describe a point—is called the dimension of the manifold. Let us denote a manifold by the letter M, and let P be one of its points.

The assignment of a Euclidean space to a neighbourhood of P is called a map. A map allows us to employ methods of mathematical analysis like calculus, as long as we stay close to P. A difference between an n-dimensional manifold M and the space $\mathbb{R}^n$ is that one map is usually not sufficient to cover the whole of M. We instead use a collection of maps called an atlas which overlap in the way that maps in an ordinary road atlas do. For M to be a manifold, these overlapping regions on various maps must smoothly patch together in the way I shall now describe. Let us consider a two-dimensional sphere, which is an example of a two-dimensional manifold. Recall the Mercator map in Figure 29. It distorts distances, but we are not worried about these at the moment. The problem with this map is that both the north and the south poles are not represented by points, but have been stretched to horizontal lines at the top and bottom edges of the map: this map is not a one-to-one projection of the sphere to the plane. We shall instead construct an atlas which consists of two maps.

The first map will be constructed by Alice, who positions herself on the north pole, N, and uses stereographic projection (Figure 48) to map all the points of the sphere except N onto the plane. To do that, she joins any point P on the sphere to the north pole by a straight line in $\mathbb{R}^3$. This line intersects the plane passing through the equator at some point P', and this P' is the image of P on Alice's map. Every point on S^2 apart from N is mapped to some point on the plane in a one-to-one fashion; for example, the image of the south pole is the origin of the plane, and the whole of the Antarctic is mapped onto a tiny region around this origin. On the other hand, the Arctic circle is mapped to a huge region of infinite area. A region covered by a small coin close to the north pole would appear far larger than the whole of Australia on Alice's map.

If the equation of the sphere is $x^2 + y^2 + z^2 = 1$ and the coordinates of N are $(0, 0, 1)$, then any point on the map constructed by Alice has coordinates (u, v), where

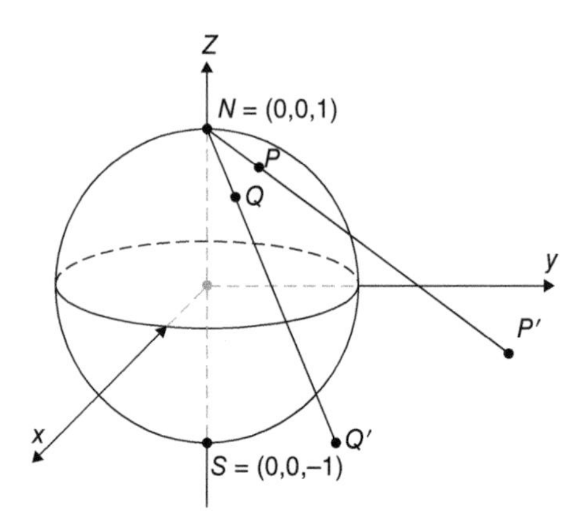

48. Stereographic projection. Two points, *P* and *Q*, are close to each other on the sphere, but their images, *P*′ and *Q*′, are far apart on the plane.

$$u = \frac{x}{1-z}, \quad v = \frac{y}{1-z},$$

which can be worked out using similar triangles. As Alice has excluded the north pole, she does not have to worry about division by zero in these formulae, as the coordinate z is never equal to one away from N. As z gets closer to 1, the coordinates (u, v) become huge, resulting in the large Euclidean distance between the projected point and the origin of the plane.

The second map is constructed by Bob, who stands on the south pole, S, and stereographically projects the sphere onto the plane from S. The coordinates of a point on Bob's map are $(\tilde{u}, \tilde{v})$, where

$$\tilde{u} = \frac{x}{1+z}, \quad \tilde{v} = \frac{y}{1+z}.$$

Alice and Bob would like to compare their maps. All points on the sphere apart from N and S have their unique images on each map. Using the defining equation of the sphere, Alice and Bob find the patching conditions

$$\tilde{u} = \frac{u}{u^2 + v^2}, \quad \tilde{v} = \frac{v}{u^2 + v^2}.$$

These expressions are well defined, as u and v cannot both be zero in the overlap region with N and S removed. From the point of view of mathematical analysis, the patching conditions are two functions from $\mathbb{R}^2$ to $\mathbb{R}^2$ which are smooth, meaning that they can be differentiated with respect to u and v. Manifolds resulting from patching relations with this property are also called smooth.

The stereographic projection and the overlap patching conditions readily generalize to spheres of any dimension, so they are all manifolds with atlases consisting of two maps, and calculus can be employed. It turns out that aside from the stereographic projection, all choices of maps and patching relations lead to the same notion of calculus on the two-dimensional sphere: there is only one way in which it can be turned into a smooth manifold. Until the middle of the 20th century, this was generally believed to be the case for spheres of all dimensions. It thus came as a surprise when in 1956 the American geometer John Milnor showed that there are several non-equivalent atlases that one can introduce on the seven-dimensional sphere: the one coming from stereographic projection, and others now dubbed the exotic spheres. In the 1980s it was shown that even the relatively familiar Euclidean space $\mathbb{R}^4$ can be turned into a smooth manifold in more than one, and in fact infinitely many, ways. Four dimensions appear to be special from the manifold theory perspective. One of the open problems in the field is to determine whether the four-dimensional sphere has any exotic structures, apart from the one coming from the stereographic projection.

According to Einstein's theory of gravitation, our Universe is a four-dimensional manifold. We shall devote the whole of Chapter 7 to studying its properties.

Riemannian geometry

While the notion of a distance function can be defined for general manifolds, it is a complicated object. An alternative was proposed by the German mathematician Bernhard Riemann in the 'Habilitation' lecture he gave in 1854. Riemann, who was a student of Gauss, adopted an infinitesimal approach based on the idea that in a sufficiently small region geometry is nearly flat, and generalized the Gaussian curvature to manifolds of more than two dimensions. The following description avoids calculus, but requires some knowledge of matrices.

Consider an infinitesimal displacement from a point *P* to a nearby point *Q*. This is given by a vector tangent to the manifold at the point *P*, represented by an arrow (Figure 49). The collection of all tangent vectors at *P* is the familiar space $\mathbb{R}^n$, where n is the dimension of the manifold. A vector $\mathbf{v}$ in $\mathbb{R}^n$ has n components $(v_1, v_2, \ldots, v_n)$ and a length, denoted by $|\mathbf{v}|$, with a square given by

$$|\mathbf{v}|^2 = (v_1)^2 + (v_2)^2 + \ldots + (v_n)^2.$$

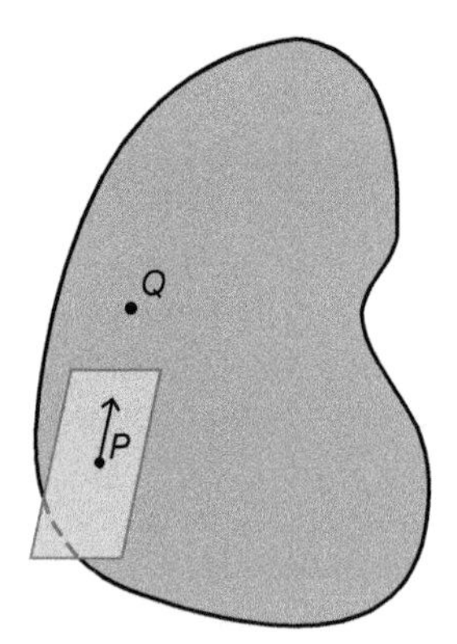

49. Tangent vector on the tangent plane.

In our case each tangent space is equipped with its notion of length, given by an n by n symmetric matrix which we shall call $\mathbf{g}_P$ to indicate its dependence on the point P. The squared length of a vector tangent at P is then given by $\mathbf{v}\, \mathbf{g}_P\, \mathbf{v}^T$. If $\mathbf{g}_P$ is the identity matrix, then this squared length equals the usual Euclidean one. A Riemannian metric is a collection of such n by n matrices, with one attached to each point P on the manifold, and varying smoothly with P: if P is close to Q, then the difference between the matrices $\mathbf{g}$(P) and $\mathbf{g}$(Q) is small. This metric is assumed to be positive definite in the sense that $\mathbf{v}\, \mathbf{g}_P\, \mathbf{v}^T \geq 0$, with equality only if all components of the vector $\mathbf{v}$ vanish. Given a Riemannian metric, the distance between any two points along a given curve can be computed by picking a large number of points along the curve and dividing the curve into segments. Each of these points is connected to a neighbouring point by a vector tangent to the curve (Figure 50). We now compute the length of each of these vectors using the Riemannian metric and add these lengths together. When the number of the segments tends to infinity, the length of each of the vectors tends to zero but the sum of the lengths is a finite number—the length of the curve. To make this process rigorous

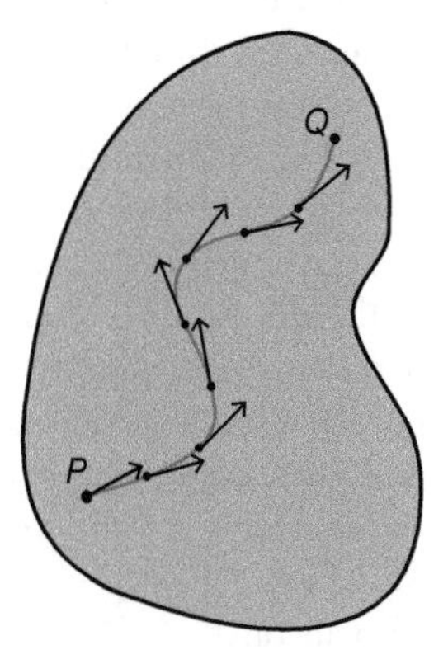

50. Length of a curve.

and amenable to computations, calculus needs to be invoked, where summation is replaced by integration.

The curved manifold does not need to be embedded in some higher-dimensional flat space in order for all this to make sense, but it is easier to draw pictures if it is. However, it turns out that any Riemannian manifold can, in fact, be embedded in a Euclidean space of a sufficiently high dimension in such a way that the Riemannian metric is induced by the Euclidean distance. This remarkable result was proven in 1966 by John Nash, the American mathematician who was later awarded a Nobel Prize in Economics for his work on game theory. According to Nash's theorem, the maximal number of dimensions of the flat ambient space needed to embed an n-dimensional Riemannian manifold is $\frac{1}{2}n(n+1)(3n+11)$.

The Riemannian metric is a collection of $n(n+1)/2$ numbers associated with each point of a manifold. The methods developed by Riemann use calculus to derive concepts like distance, area, and angles from the Riemannian metric. I shall now describe, without reverting to calculus, how one can make sense of the curvature of Riemannian manifolds. Having defined the length of the curve, we can attempt to pick the shortest curve, called the geodesic, connecting any two points P and Q on the manifold. At a first glance, this appears to be a formidable task: to make progress, one needs to take all possible curves between P and Q into consideration, compute the distances $d(P, Q)$ along all these curves, and seek the optimal one, assuming that it even exists! Fortunately, there is a tool, known as the calculus of variations, which does all that with relative ease as long as the Riemannian metric has been specified. This comes with a word of warning, as the 'shortest' curve between two points does not have to exist at all. Consider, for example, a 'punctured' Euclidean plane $\mathbb{R}^2$, where the origin (0, 0) has been removed, and try to connect two points P and Q on the x-axis on the opposite sides of the puncture. Given any curve joining P and Q, a shorter curve may be constructed by zooming nearer to the puncture, but the limiting shortest curve does not exist.

Geodesics do, however, exist locally, in a neighbourhood of any point on the manifold. So, in the above example, we can connect P with Q by a geodesic (which is just a straight line in this case), as long as Q is sufficiently close to P. The precise statement is that, given a point P on M and a direction vector at P, there exists a unique geodesic through P in this direction. Geodesics on the plane are straight lines. On the sphere they are arcs of great circles, but in this case there can be more than one geodesic between two points: the north and the south poles are connected by infinitely many geodesics. Geodesics on the Poincaré disc are the hyperbolic lines discussed in Chapter 3.

Two-dimensional Riemannian manifolds are just surfaces, and their Riemann curvature is the Gaussian curvature I explained earlier in the chapter. For higher-dimensional manifolds, the curvature at any point is no longer a single number but instead an array of numbers assembled into what is called a tensor. To obtain these numbers, pick two independent directions in the tangent space to M at P. These directions span a plane, and the geodesics emanating in each direction on this plane lie, at least near the point P, on a two-dimensional surface Σ. We compute the Gaussian curvature of Σ using the distance function resulting from restricting the Riemannian metric from M to Σ. This process is then repeated for all two-dimensional planes and the corresponding surfaces. For a general Riemannian manifold, the resulting Gaussian curvatures will be different and will depend on a choice of the plane. They are assembled into the Riemann curvature. For some manifolds, like the sphere, the curvature is positive and the same at every point, but for general manifolds it varies from point to point. The sign of the curvature has an effect on the behaviour of neighbouring geodesics. If the curvature is negative, then two close-by geodesics in parallel directions move away from each other. The positive curvature has the opposite, focusing effect: imagine two meridians emanating from the equator in parallel directions pointing out to the north pole. They converge to each other, eventually meeting at the north and south poles. This curvature effect has striking consequences, and underlies the Penrose–Hawking singularity theorems stating that the presence of mass leads to space-time singularities and black holes. We shall return to this point in Chapter 7.

Chapter 5
Projective geometry

Draw two parallel lines on a large piece of paper. Lay the paper on a flat surface, and look at it from a distance. The lines still look straight, but appear to come together. For a more striking example, take a drive down Highway 46 east from Gackle in North Dakota. The road does not bend for about 30 miles, and its boundaries will appear to intersect at an *infinitely far-away point* on the horizon.

In projective geometry, such *points at infinity* are added to the Euclidean plane, and are regarded on an equal footing with all other points. In Euclidean geometry, two lines intersect at a unique point unless they are parallel. From the projective perspective, these 'parallel lines' intersect at one of the points at infinity. Experimenting with three pairs of parallel lines in perspective suggests that the corresponding intersection points all lie on a line. In fact, all points at infinity lie on one line—which is therefore called *the line at infinity*—and our first definition of the projective plane will be as a Euclidean plane with one line adjoining it. In this chapter we shall explore the geometry of this plane.

Projective plane and the line at infinity

Compare the following two paintings (Figure 51).

51. **(a) Duccio di Buoninsegna (circa 1310), *The Capture of Jesus.* (b) Canaletto (circa 1730), *Interior Court of the Doge's Palace***

The medieval painting of di Buoninsegna represents people, objects, and space as they *are*, paying attention to Euclidean proportions and sizes. The Renaissance painting of Canaletto depicts the three-dimensional space as it *looks*. The Euclidean geometric proportions have been distorted: the Basilica San Marco on the far side of the square is in reality taller than the arch at the front of the painting, and yet it has been presented as smaller using the concept of perspective. The discovery of perspective led the Renaissance artists to seek geometric schemes that enabled them to represent three-dimensional space filled with objects of various sizes and relative distances on a two-dimensional canvas. The key concept underlying the perspective drawing is the projection. I shall assume that the idealized artist has only one eye, represented by a point. The canvas used by the artist is a glass screen put between the scene and the eye. Each point in the scene painted by the artist is connected to his eye by a ray of light, which intersects the screen at some other point. The set of points arising in that way on the screen forms a plane section, which is then turned into a painting (Figure 52). Fixing the position of the artist's eye, but changing—for example by a translation and a rotation—the configuration of the glass screen results in a different painting, and the problem for a mathematician is to determine a relation between the two projections.

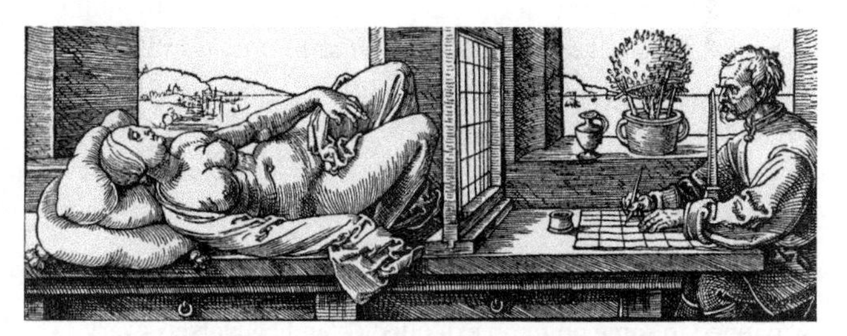

52. A perspective machine, from Albrecht Dürer's *Painter's manual* (1525).

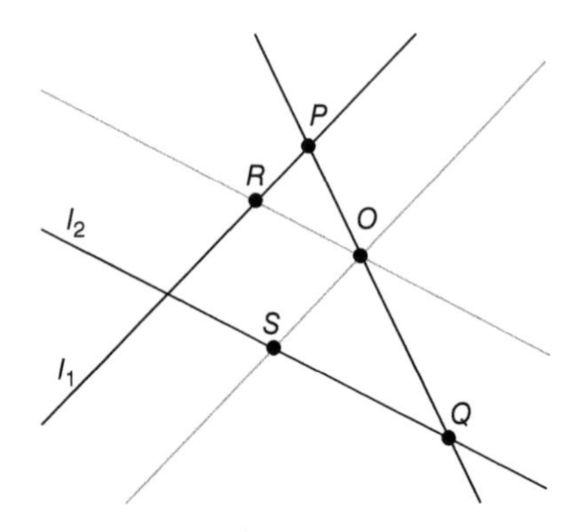

53. Projection from a line to a line.

To gain more insight into what is going on, I shall focus on the simplest example, which is a projection of one line in the Euclidean plane to another such line from a point O which does not belong to either of the two lines. Let us call the lines l_1 and l_2. The point P on the line l_1 is projected to the point Q, which is the intersection of the line l_2 with the line containing the segment OP (Figure 53).

There are two problems with this projection. First, if R is the point on l_1 such that the segment OR is parallel to l_2, then R is not projected to any point on l_2. This problem can be fixed by adding an additional point to l_2—call it *a point at infinity*—and declaring it to be the image of R under the projection. Second, if S is the point on l_2 such that OS is parallel to the line l_1, then there is no point on l_1 which projects to S. We use a similar fix: extend the line l_1 by attaching a point at infinity to it, and declare that this point projects to S. This way we have made the projection what mathematicians call a bijection: every point on the extended line l_1 is mapped to some point on extended l_2, and conversely every point on extended l_2 is a projection of some point on extended l_1. These extended lines are called projective lines. A line on the Euclidean plane is denoted by $\mathbb{R}^1$, while a projective line is denoted by $\mathbb{P}^1$. The relation between the two is

$$\mathbb{P}^1 = \mathbb{R}^1 + \{\infty\},$$

where $\{\infty\}$ is the notation for the point we attached to each line.

Now I turn to the more realistic projection from point O of the plane Π_1 in Euclidean space $\mathbb{R}^3$ to another plane, Π_2, which is not parallel to Π_1. We assume that O does not belong to any of the two planes, and, as before, define the projection of any point P on Π_1 to be the intersection of the line containing points P and O with the plane Π_2. This time there is a whole line l on Π_1 which does not project to any region of Π_2. This is the line of points which lie on the plane containing O, and is parallel to the plane Π_2. Similarly, there is a line on Π_2 to which no points project (Figure 54).

We address both problems by extending the planes Π_1 and Π_2 to projective planes: a projective plane $\mathbb{P}^2$ is the Euclidean plane together with a projective line attached 'at infinity':

$$\mathbb{P}^2 = \mathbb{R}^2 + \mathbb{P}^1.$$

This line at infinity is the vanishing line on an artist's canvas. It is also the line that motorists see on the horizon, where all the sides of straight roads intersect.

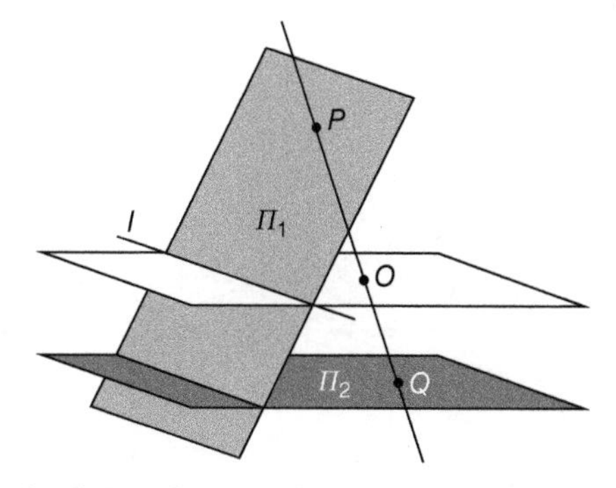

54. Projection from a plane to a plane.

The Desargues theorem

Changing the relative positions of planes Π_1 and Π_2 in space results in a different projection. This, in perspective drawing, leads to a different pictorial realization of the same reality, with the glass screen between the artist's eye and the space repositioned. Distances and shapes of objects differ for both projections. For example, a square projects onto a square if the planes Π_2 and Π_1 are parallel, but it will project onto some other quadrilateral if they are not. Some properties do carry over from one projection to another. A straight line will remain straight, a quadrilateral will not turn into a triangle, and two lines intersecting at a point will still intersect, albeit possibly at infinity.

The first formal study of the properties shared by all projections was completed by the 17th-century French architect and engineer Girard Desargues. One of his findings concerns the relationship between two projections of the same triangle (Figure 55).

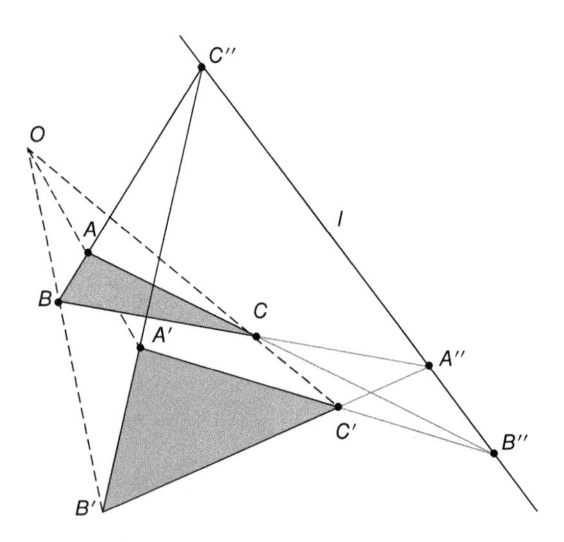

55. Desargues's theorem.

The two triangles ΔABC and $\Delta A'B'C'$ are in perspective, which means that they are related by a projection from point O not belonging to either of the two triangles. Thus the lines through the corresponding vertices A and A', B and B', C and C' all pass through O.

> **Desargues's theorem.** If two triangles are in perspective, then the points of intersections of corresponding sides lie on a single line.

The proof of Desargues's theorem is given in two separate cases: when the triangles ΔABC and $\Delta A'B'C'$ do not lie on the same plane, and when they do. The first case of non-coplanar triangles is more general. I shall split its proof into five steps. In this proof, and in the rest of this chapter, the line (which is necessarily unique) containing two points A and B will be referred to as AB. There is no risk of confusion, as neither 'line segments' nor 'orientation' appear in projective geometry.

Step 1 The planes containing the triangles ΔABC and $\Delta A'B'C'$ intersect in a line (possibly at infinity) which I shall call l.

Step 2 The two lines OA and OB determine the unique plane Π, which contains both lines, as well as points A' and B'. This plane also contains the lines AB and $A'B'$, which intersect at a point C'' on the plane Π (this point will be at infinity of Π if the lines AB and $A'B'$ are parallel).

Step 3 The point C'' is in the plane of both triangles ΔABC and $\Delta A'B'C'$; therefore, C'' belongs to the line l.

Step 4 Apply Steps 2 and 3 to the lines OA and OC. They intersect at a point B'' on the line l. Similarly, the lines OB and OC intersect at A'' on the line l.

Step 5 Thus A'', B'', and C'' all belong to the same line.

Homogeneous coordinates

A drawback in our discussion of the projective plane is that the line at infinity does not appear to be on an equal footing with other lines. I shall remedy this by giving another definition of the projective plane which treats all lines in a more democratic fashion, and will then prove that the two definitions agree.

The projective plane is the space of all lines in Euclidean three-space $\mathbb{R}^3$ which pass through the origin. This definition links directly with perspective drawing: the origin in $\mathbb{R}^3$ is the position of the artist's eye. Any point in space is linked to the artist's eye by a light ray, which is a straight line through the origin. But all points on the ray will be seen as one by the artist. Thus two points with coordinates $(1, 2, 7)$ and $(2, 4, 14)$ are identified, as they lie on the same line through $(0, 0, 0)$. A line not through the origin—this may be the side of a straight road the artist is painting—is contained in a unique plane through the origin (Figure 56). This line corresponds to a line on the projective plane of the canvas, and conversely lines in the projective plane correspond to planes through the origin in $\mathbb{R}^3$. We can use this to prove the important property of the projective plane:

P1 Two lines on the projective plane intersect at a point.

Indeed, two planes through the origin intersect in a unique line through the origin, but this line by definition corresponds to a

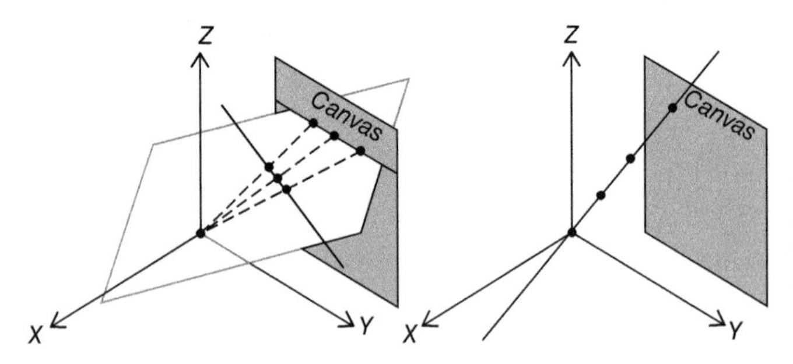

56. Points and lines on a projective plane.

point on the projective plane $\mathbb{P}^2$. Note that the result we have just established does not hold in Euclidean geometry, where one would have to exclude parallel lines from the statement: **P1** violates Euclid's axiom **E5'**. The notion of parallelism does not exist in projective geometry, and all pairs of lines are on an equal footing.

We need to reconcile this new definition of the projective plane with the previous picture, involving adding a line at infinity to the Euclidean plane. To do it we introduce the coordinate system X, Y, Z on $\mathbb{R}^3$. I am deliberately using capital letters to distinguish these coordinates from the Cartesian coordinates on the plane. Our definition of the projective plane identifies $[X, Y, Z]$ with $[\lambda X, \lambda Y, \lambda Z]$, where λ is any non-zero number. Once this identification is made, the coordinates $[X, Y, Z]$, not all of which are simultaneously zero, label points on $\mathbb{P}^2$. They are called the homogeneous coordinates on the projective plane. We can now recover the Euclidean plane together with an adjoining line at infinity as follows: assume that $Z \neq 0$. Then $[X, Y, Z]$ is identified with $[X/Z, Y/Z, 1]$, and by setting

$$x = \frac{X}{Z}, \quad y = \frac{Y}{Z}$$

we have recovered a Euclidean plane $\mathbb{R}^2$ with coordinates (x, y). We have so far excluded all points on the projective plane given by $[X, Y, 0]$. To account for these points, first assume that $Y \neq 0$, in which case $[X, Y, 0]$ can be identified with $[X/Y, 1, 0]$. This is a line $\mathbb{R}$ with the Cartesian coordinate $w = X/Y$. We are still not done, as both Y and Z could have been zero, which leaves us with $[X, 0, 0]$. This time X is definitely not zero, so we can identify this with $[1, 0, 0]$, which is a point on the projective plane. We have therefore proved that

$$\mathbb{P}^2 = \mathbb{R}^2 + \mathbb{R}^1 + \{\text{point}\}.$$

The last two factors—the Euclidean line, and a point—combine to give the projective line $\mathbb{P}$, thus reconciling the two definitions of the projective plane.

Conics

We saw in Chapter 2 that not all geometric constructions can be achieved using a compass and ruler or their geometric counterparts—lines and circles. Some of these constructions are made possible if other types of curves are allowed. For example, doubling a cube, which comes down to constructing the cube root of two, can be accomplished by employing parabolas. The two points of intersection of the parabolas,

$$y = x^2 \quad \text{and} \quad y^2 = 2x,$$

can be found by squaring the first equation and substituting into the second. This gives $x^4 = 2x$, which is solved by $x = 0$ and $x = \sqrt[3]{2}$.

Parabolas are examples of conics. A conic is a curve of intersection of a cone with a plane. I shall call the vertex of the cone O, and the plane defining the conic Π. I shall also exclude planes which contain the vertex of the cone. Three possible types of conics (Figure 57) can be distinguished by examining the plane Π_1, which is parallel to Π and passes through the vertex O.

- If Π_1 intersects the cone in two lines, then the conic is a hyperbola.
- If Π_1 is tangent to the cone, then the conic is a parabola.
- If Π_1 only intersects the cone at its vertex, then the conic is an ellipse.

These types of conics were studied in detail by the Greeks in the 2nd century BC, primarily by Apollonius of Perga who covered the whole theory of conics in eight volumes. This contains the 'goat eating grass' definition of ellipse. If you want to cut an ellipse in your lawn, stick two pegs into the ground, and connect them with a loose but not extendable rope. Attach a hungry goat to the rope

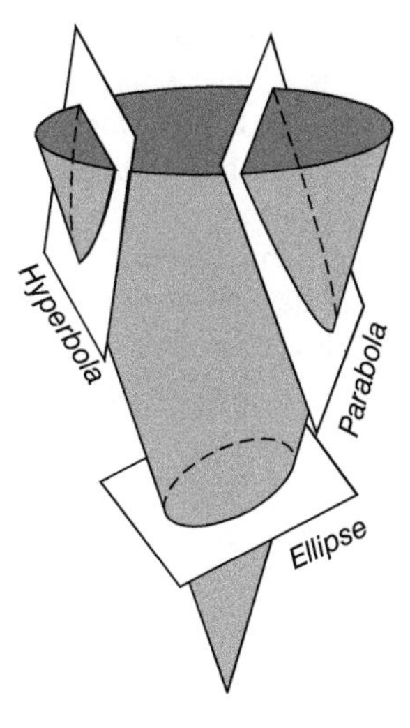

57. Conic sections.

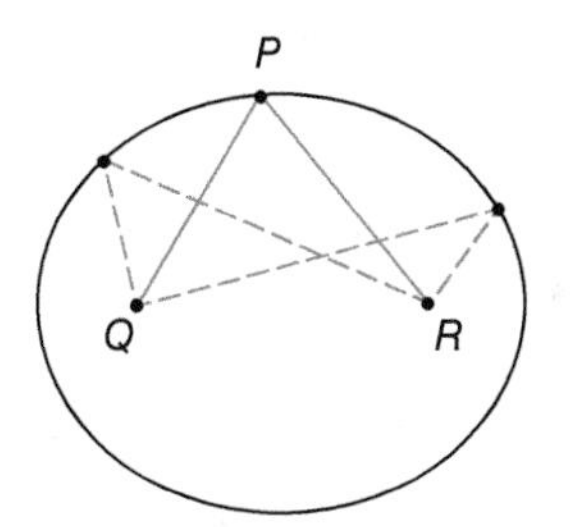

58. The goat definition of the ellipse.

so that it can move freely along the rope. The boundary of the region eaten by the goat is an ellipse. This reflects a definition which states that the ellipse is a locus of all points P on a plane such that the sum of distances $|PQ|$ and $|PR|$ from P to two fixed points Q and R is constant (Figure 58).

The goat definition of the ellipse is pictorial and accessible (not least for the goat), but it relies on the concepts of Euclidean geometry such as the total length of the rope. In fact, the natural arena for studying conics is the projective plane. In projective geometry all conics are the same, as the plane of any conic can be projected onto the plane of any other conic from the vertex of the cone. Thus every conic is the projection of a circle to the plane of this conic, and all possible conics arise as such projections. The difference between a hyperbola and an ellipse reflects a different choice of canvas.

To give an example of a coordinate expression of a conic, take the cone centred at the origin in $\mathbb{R}^3$, given by an equation

$$x^2 + y^2 = z^2,$$

and intersect it with the plane $5z - 3x - 20 = 0$. Substituting z from the second equation into the first gives

$$16x^2 - 120x + 25y^2 - 400 = 0.$$

This is an equation of the ellipse, as can be seen by rewriting it as

$$\left(\frac{4}{5}x - 3\right)^2 + y^2 = 25.$$

The general equation of a conic is

$$ax^2 + by^2 + 2cxy + 2dx + 2ey + f = 0,$$

where a, b, c, d, e, f are constant parameters—in the previous example, $a = 16, b = 25, c = 0, d = -60, e = 0, f = -400$. Not all of these parameters can simultaneously be equal to zero, so the number of essential parameters is five: if, for example, $f \neq 0$, then we can divide both sides of the conic equation by f, and we are left with the ratios $a/f, b/f, c/f, d/f$, and e/f as five parameters determining the conic. This analytical approach can be used to give a simple—if somewhat tedious—proof of a fact known to the Greeks:

There is a unique conic through any given coplanar five points, no three of which lie on the same line.

Let the five points $P_1, P_2, \ldots, P_5$ have coordinates $(x_1, y_1), (x_2, y_2), \ldots, (x_5, y_5)$ on the plane of the conic. We substitute (x_1, y_1) to the equation of the conic—it has to hold, as P_1 belongs to the conic—which gives one equation for the five parameters. We can do the same with the (x, y) coordinates of the remaining four points. This gives a system of five equations whose five unknowns are the parameters of the conic. Employing algebra shows that this system has a unique solution, as long as no three of the five points are collinear. This argument also shows that in general it is not possible to draw a conic through six given points. Adding another point would result in six equations for five unknowns, where there is usually no solution.

In what circumstances can a conic be drawn through six points? Connecting the six points with line segments gives a hexagon. Therefore, an equivalent problem is to find a special property of those hexagons which can be inscribed into a conic. The answer is provided by a theorem, stated in 1639 by the French mathematician Blaise Pascal when he was only 16 years old.

Pascal's theorem. If a hexagon is inscribed in a conic, then the three points of intersections of pairs on opposite sides of the hexagon lie on the same line (Figure 59).

Pascal was a contemporary of Desargues, and their works on projective geometry shared the same fate: they were ignored by their contemporaries and largely forgotten for some 200 years. Desargues and Pascal were ahead of their time. Projective geometry was rediscovered and revived by a group of 19th-century French mathematicians, primarily Jean-Victor Poncelet. Poncelet was an officer in the French army during Napoleon's invasion of Russia. He was wounded and captured in 1812, and spent a year in

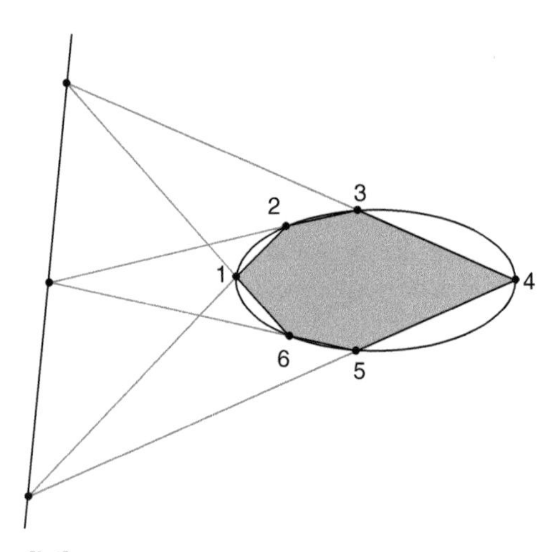

59. Pascal's theorem.

a Russian prison, where he studied projective geometry and put it in what we now regard as the modern framework.

It is to Poncelet that we owe one of the most remarkable results about conics. The Poncelet theorem is concerned with a pair of distinct conics, which can be drawn as two ellipses, one placed inside the other (Figure 60). Pick a point P_1 on the outer ellipse $\mathcal{A}$, and draw a line through this point which is tangent to the inner ellipse $\mathcal{B}$. This line will intersect the ellipse $\mathcal{A}$ at a point P_2. Now draw another tangent line to $\mathcal{B}$ through P_2. This tangent intersects $\mathcal{A}$ at another point P_3. Iterating the process gives a sequence of points on the outer ellipse $\mathcal{A}$. For a generic choice of two ellipses (Figure 60(a)), all points in this sequence are distinct, so none of the tangents will ever cross the initial point P_1. There are special pairs of conics, like the one depicted in Figure 60(b), where after a finite number of steps we return to P_1 and produce an n-gon inscribed in $\mathcal{A}$ and circumscribed about $\mathcal{B}$. In Figure 60(b) this happens after three iterations. The Poncelet theorem states that if there exists an n-gon inscribed in one conic $\mathcal{A}$ and circumscribed

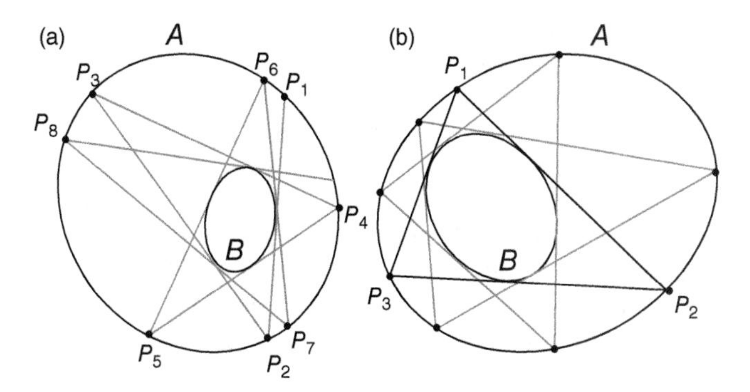

60. Poncelet's theorem.

about another conic $\mathcal{B}$, then there exist infinitely many such n-gons: for any point P on the conic $\mathcal{A}$, there exists an n-gon with P as its vertex.

Projective duality

The two properties of the projective plane we have uncovered so far are:

P1 Two lines intersect at a unique point.
P2 Two points are contained in a unique line.

The second property also holds in Euclidean geometry, but the first one does not if the lines are parallel. The projective point of view puts all lines on equal footing so that, as we have seen, **P1** is true for all lines. Moreover, the first property turns into the second property if we replace the words *lines* by *points*, and *points* by *lines*. This is an example of *projective duality*. It turns out that if such a replacement is made, not only simple statements like **P1** and **P2** remain true, but every theorem in projective geometry does too. The arena for the projective duality is the interplay between the projective plane $\mathbb{P}^2$ and the dual projective plane—we shall call it $\mathbb{P}^2{}_\star$—which is the space of all lines in $\mathbb{P}^2$ (Figure 61).

One way to visualize projective duality is to draw the projective plane and its dual, as shown in Figure 61, and replace points on $\mathbb{P}^2$ by lines on $\mathbb{P}^2{}_\star$ and vice versa in such a way that the properties **P1** and **P2** are interchanged by the duality. We can instead make the duality transparent and draw points and their dual lines on one projective plane (Figure 62). In this figure, the duality is based on the conic $\mathcal{C}$, which I have chosen to be an ellipse.

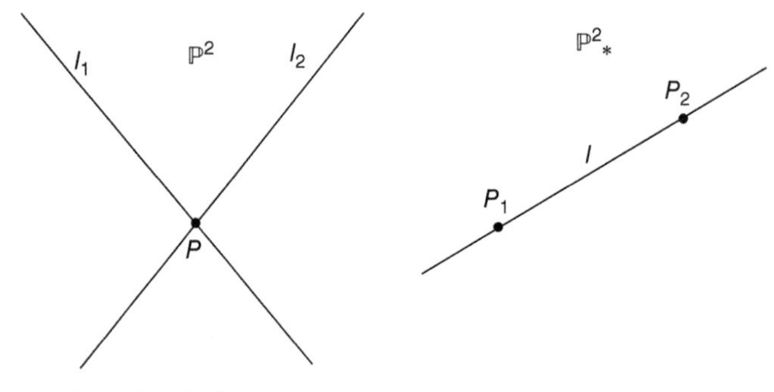

61. Projective duality.

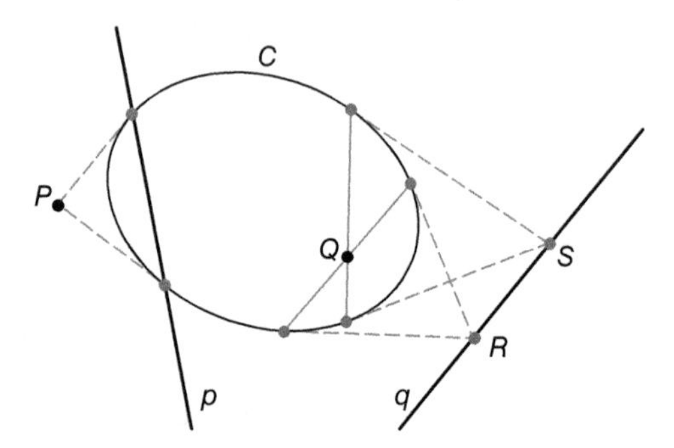

62. Duality with respect to a conic.

To find the line dual to a point P outside the ellipse, draw two lines through P which are tangent to $\mathcal{C}$. The thick line p connecting the tangency points is the dual of P. In the limiting case when P lies on the conic, the dual line is just the tangent to $\mathcal{C}$ at P. What about if the point, call this one Q, is inside the ellipse? We then take any two lines through Q, each line intersecting $\mathcal{C}$ at a pair of points. The two tangents to $\mathcal{C}$ emanating from one of the pairs intersect at some point, call it R, outside the ellipse. The tangents emanating from the second pair intersect at another point, S. By property **P1**, there is a unique line containing the points R and S. This is the dual line to Q. The concept of projective duality extends to smooth curves in the projective plane. A tangent line to any point on a curve γ in $\mathbb{P}^2$ corresponds to a point in the dual projective plane. Thus the one-parameter family of tangents to all points of γ defines a curve in $\mathbb{P}^2{}_\star$. In particular, the dual curve to a conic is also a conic.

What theorem in projective geometry arises as the statement dual to the Pascal theorem about a hexagon inscribed inside a conic $\mathcal{C}$? In the dual plane, the six vertices of the hexagon are turned into six edges of a polygon circumscribed about $\mathcal{C}$. The three collinear points of intersections in Pascal's theorem become the three diagonals of the polygon connecting the opposite edges. The statement dual to Pascal's theorem is that these three diagonals intersect at one point. This remarkable fact was discovered by the French mathematician Charles Brianchon in 1810. The proof of Brianchon's theorem follows from the principle of duality applied to Pascal's theorem. I shall use the duality based on the conic $\mathcal{C}$, and split the proof into three steps (Figure 63):

Step 1 The vertex A of the circumscribed hexagon is dual to the line a of the inscribed hexagon. This duality relation also holds for the vertices B, C, D, E, F and the lines b, c, d, e, f.

Step 2 The line connecting the vertices A and D is, by the principle of duality, dual to the intersection point of the lines a and d.

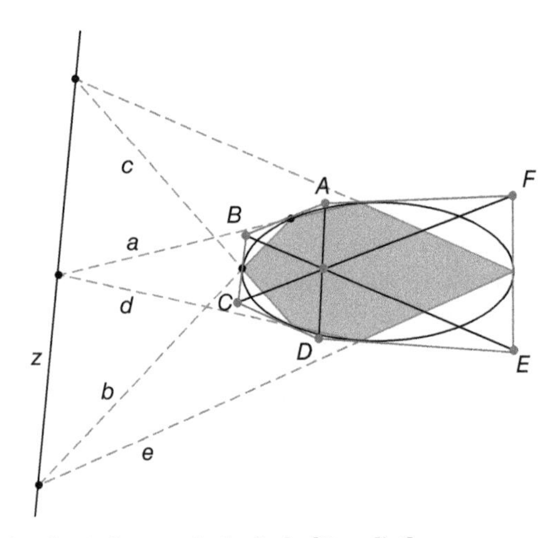

63. Brianchon's theorem is the dual of Pascal's theorem.

Step 3 The three points of intersections of pairs of lines (a, d), (c, f), and (b, e) lie on the same line z. Therefore, by the principle of duality, the three lines AD, CF, and BE intersect at the point Z. This is the point dual to the point z, and this proves the theorem.

In this picture the duality is taken with respect to the conic $\mathcal{C}$, which is unchanged by the duality operation.

While the correspondence between the projective plane and its dual can be revealed by drawing pictures, it can also be understood analytically, using the homogeneous coordinates. If $[X, Y, Z]$ are the homogeneous coordinates of a point in $\mathbb{P}^2$, then an equation of a line in $\mathbb{P}^2$ (or equivalently, a plane through the origin in $\mathbb{R}^3$) is

$$AX + BY + CZ = 0.$$

The numbers $[A, B, C]$, not all of which can be zero, are coordinates on the dual projective plane ${\mathbb{P}^2}_\star$. The projective duality is based on interpreting this equation in two ways: fixing a point with coordinates $[A, B, C]$ in $\mathbb{P}^2{}_\star$ gives a line in $\mathbb{P}^2$, as we have already seen. We can instead fix a point with $[X, Y, Z]$ in $\mathbb{P}^2$. This gives a line in $\mathbb{P}^2{}_\star$.

Projective geometry in the 21st century

Interest in projective geometry came in waves. The first wave was the work on geometric perspective, formalized by the artists Leonardo da Vinci, Piero della Francesca, and Albrecht Dürer. Then, a hundred years later came the second wave, following the theorems of Pascal and Desargues. These were then forgotten for two centuries, and revived by Poncelet and his followers in the 19th and 20th centuries. Research in projective geometry declined in the second half of the 20th century, and the subject has all but disappeared from schools.

However, the last three decades have seen a renaissance of projective geometry in both applied and pure forms. An example of the former is the problem of reconstructing a three-dimensional scene from a sequence of two-dimensional images in computer graphics. Rather than capturing the space through its projection onto a plane, one aims to recover the spatial proportions and distances accurately from the knowledge of several projections. The mathematical details and algorithms used here are based on comparing the homogeneous coordinates of two or more projective planes. It is likely that the demands of virtual reality software will lead to more interest in projective geometry.

In pure mathematics, research has focused on projective differential geometry, where properties of curves and surfaces (generalizing projective lines and planes) are studied on general manifolds, which we encountered in Chapter 4. From the point of view of this geometry, the Euclidean plane, the Poincaré disc, and the round sphere are all equivalent, as there exist projections mapping the geodesics of any of these geometries to the geodesics of the other two. These three geometries have constant Gaussian curvature—positive for the sphere, negative for the disc, and zero for the plane. It is not possible to find a projection of a general curved surface with non-constant Gaussian curvature onto a plane in a way which maps its geodesics to straight lines. Until recently it was not known how to determine whether a surface covered with curves, with one curve through each point and in each direction, can be projected onto another surface such that the curves are mapped to geodesics minimizing some distance function. This question, known as the metrizability problem in projective geometry, was posed by the French mathematician Roger Liouville, who, in 1889, explained how to encode a given family of curves into a function f of three variables. The problem was solved in 2009, and the solution relied heavily on symbolic computer algebra only developed in the 21st century: the criteria for metrizability reduce to the vanishing of algebraic expressions which contain over 2,000 terms involving the function f, and its derivatives up to order eight. The problem of Liouville can be generalized from surfaces to manifolds of more than two dimensions, where the solution is still not known.

Chapter 6
Other geometries

Carl Friedrich Gauss, whom we encountered in Chapter 4 in our discussion of the Gaussian curvature, made important contributions outside geometry, in just about every branch of mathematics: the Gauss lemma in number theory and Gaussian distributions in statistics, to name only two. Such a broad perspective from one mathematician has not been seen since the second half of the 20th century. It may be that we are just lacking geniuses of Gauss's calibre, but a more likely explanation is that mathematics has increasingly become too specialized for anyone to grasp it as a whole. Thousands of PhD students take four years to complete research projects generalizing research projects completed by their supervisors 20 years earlier. While this has led to some truly remarkable progress, it came at the price of introducing narrow specialization, which makes it difficult for mathematicians to share their findings with non-specialists, or even among themselves.

Geometry, which has been one of the fastest-developing areas of mathematics, has not been immune to this. The Atiyah–Singer index theorem, proved in 1963 by Michael Atiyah and Isadore Singer, is one of the most celebrated results in geometry in the last 60 years. It *roughly* states that for an elliptic differential operator on a compact manifold, the analytical index related to the dimension of the kernel of the operator is equal to the topological index. Let us not worry about the proof of this theorem, and focus only on its statement. To make sense of it, you need to know the definition of compact manifolds (a final-year Differential Geometry course at Cambridge), elliptic operators and their kernels (a second-year Analysis course, followed by a final-year course in Partial Differential Equations), and topological indices, which are not taught at an undergraduate level. This means that the best students who graduate with a mathematics degree from Cambridge are not capable of even understanding the statement of the Atiyah–Singer theorem. Note that this theorem was proved more than half a century ago—there have been hundreds of remarkable results established in geometry since then, which are accessible to only a handful of people. A researcher in algebraic geometry will struggle to follow a differential geometry lecture. Differential geometers will find it hard to make sense of non-commutative geometry, and so on.

This chapter offers an overview of some of these modern geometries and their links with other areas of mathematics.

Erlangen programme

Two geometric shapes are congruent in Euclidean geometry if one of them can be moved into the other by translating, rotating, and reflecting it in space. For example, all squares of side one on the plane are congruent, and once we understand geometric properties of one of these squares—such as the distance between the opposite corners, or the area of an inscribed circle—we will not need to repeat the computations for any of the other squares. All such squares are the same, up to the congruence, and so are their properties. The programme I shall now describe reverses this process. Specifying a class of allowed transformations—the rotations and translations for Euclidean geometry—allows us to define the geometry itself.

Consider a transformation of the $\mathbb{R}^2$ plane mapping a point P with coordinates (x, y) to a point $\bar{P}$ with coordinates $(\bar{x}, \bar{y})$ given by

$$\textbf{Aff} \qquad \bar{x} = ax + by + c, \quad \bar{y} = dx + ey + f,$$

where the parameters of the transformation (a, b, c, d, e, f) are fixed real numbers such that $ae - bd \neq 0$. Such a transformation is called *affine*. The image of a straight line under an affine transformation is another straight line, and the order of points on the lines is preserved. The distance between points, however, is not preserved. To see it, take two points P and Q with coordinates $(0, 0)$ and $(0, 1)$, respectively, so that the distance $d(P, Q) = 1$, and consider an affine transformation with $a = 1, b = 0, c = 1, d = 0$, $e = 2, f = 0$. We compute the coordinates of the transformed points $\bar{P}$ and $\bar{Q}$ to be $(1, 0)$ and $(1, 2)$, so that the distance $d(\bar{P}, \bar{Q}) = \sqrt{(1-1)^2 + (2-0)^2} = 2$.

The *affine geometry* is the study of those properties of geometric shapes which do not change under affine transformations. Affine transformations do not preserve distances or angles, but they map lines to lines and preserve parallelism, centroids of polygons, and the ratios of distances along parallel lines. So while a rectangle is not a well-defined object in affine geometry, a parallelogram is. Circles are mapped to ellipses by affine transformations, so conic sections are affine concepts, and, moreover, ellipses are distinguished from hyperbolas.

One of the theorems in Euclidean geometry that remains true in affine geometry is the Ceva theorem (Figure 64), named after the 17th-century Italian mathematician Giovanni Ceva. The theorem was, however, already known to the 11th-century Arab mathematician and king of Zaragoza, Yusuf al-Mutaman ibn Hud. The theorem states the following.

Ceva's theorem. Let P, Q, and R be the points on the three sides BC, CA, and AB of a triangle ΔABC. Then the relation

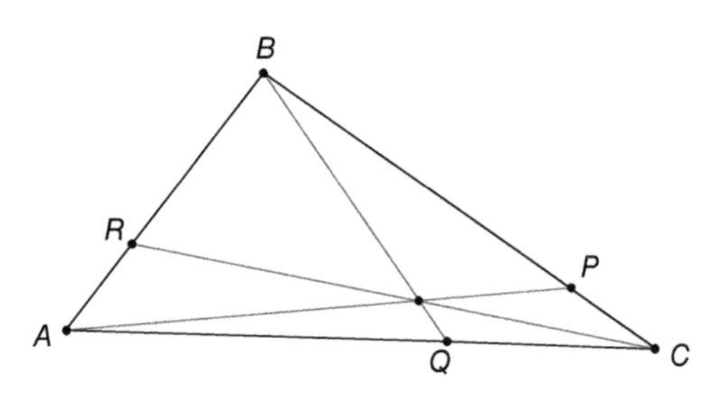

64. Ceva theorem.

holds if and only if the three lines AP, BQ, and CR intersect at one point, or are parallel.

$$\frac{|BP|}{|PC|} \cdot \frac{|CQ|}{|QA|} \cdot \frac{|AR|}{|RB|} = 1$$

Although the notion of the distance, for example $|BP|$ between the points B and P, appears in the statement of the theorem, the distances only enter as ratios. An affine transformation of the points A, B, C results in a triangle with different area and angles, and yet if the same affine transformation is applied to the points P, Q, and R, then the theorem still holds.

If we take P, Q, and R to be the midpoints of the sides of the triangle, then $|BP| = |PC|, |CQ| = |QA|$, and $|AR| = |RB|$, which makes all the ratios in the Ceva formula equal to one. We have therefore recovered a result which some readers may recall from school geometry: if a median in a triangle is defined to be a line segment joining a vertex to the midpoint of the opposite side, then, in any triangle, the three medians intersect at one point.

A composition of two affine transformations is another affine transformation. This can be seen by applying the affine transformation **Aff** with parameters $(a_1, b_1, c_1, d_1, e_1, f_1)$, followed by another such transformation with parameters $(a_2, b_2, c_2, d_2, e_2, f_2)$.

The resulting transformation will still be of the form **Aff**. A useful algebraic shortcut is to define a column vector **v** with components (x, y) and rewrite the affine transformation as $\bar{\mathbf{v}}$=A**v**+**t**, where

$$A = \begin{pmatrix} a & b \\ d & e \end{pmatrix}$$

is a 2 by 2 matrix and **t** is a column vector with components (c, f). A composition of two affine transformations $(A_1, \mathbf{t}_1)$ followed by $(A_2, \mathbf{t}_2)$ is the affine transformation $(A_2 A_1, A_2\, \mathbf{t}_1 + \mathbf{t}_2)$. Moreover, any affine transformation has a unique inverse such that the composition of the transformation with its inverse is the identity transformation given by $a = e = 1, b = c = d = f = 0$. Therefore, it follows that the collection of all affine transformations forms a group. As the concept of a group plays a central role in our discussion, it is worth digressing to define it properly. A group is a set G, together with an operation $\star : G \times G \to G$ that assigns an element $\mathbf{h} \star \mathbf{g}$ of G to any two elements **h** and **g**, and such that the following three conditions hold:

G1 There exists an element $\mathbf{e} \in G$, called the identity, such that

$$\mathbf{g} \star \mathbf{e} = \mathbf{e} \star \mathbf{g},$$

for all $\mathbf{g} \in G$.

G2 For any $g \in G$, there exists an inverse element $\mathbf{g}^{-1}$ such that

$$\mathbf{g} \star \mathbf{g}^{-1} = \mathbf{g}^{-1} \star \mathbf{g} = \mathbf{e}.$$

G3 For any elements **h**, **g**, **k** in G,

$$\mathbf{h} \star (\mathbf{g} \star \mathbf{k}) = (\mathbf{h} \star \mathbf{g}) \star \mathbf{k}.$$

The third axiom **G3** is called associativity. It need not concern us for the groups of transformation where the product $\star$ is defined by compositions of transformations. All transformations are associative.

If $\mathbf{g} \star \mathbf{h} = \mathbf{h} \star \mathbf{g}$ for all elements of G, then the group G is called Abelian. An example of an Abelian group of transformations is the group consisting only of translations of a plane. Its identity element is the translation by the zero vector. The affine group **Aff** is not Abelian, as the matrix product is non-commutative.

The German mathematician Felix Klein had the idea of *defining* a geometry as a pair consisting of a space S (in our case S is the two-dimensional plane) and a group G (in our case this G is the group of all affine transformations of the plane) transforming points of S to other points of S. The subject of such geometry is the study of properties of shapes which do not change under the group transformations. In particular, the same space S can be a playground for different geometries if different groups are chosen. Klein published his work in 1872 when he worked in the city of Erlangen in Bavaria, and his programme, active to this day, is named after this city.

Let us see how the Euclidean geometry on a plane $\mathbb{R}^2$ fits into this scheme. The underlying group consists of isometries: the transformations preserving the Pythagorean distance between points. The isometries of a plane are rotations, translations, and reflections. They are all special cases of the affine transformations **Aff**. An anticlockwise rotation by an angle θ around the origin corresponds to the transformation **Aff** with

$$a = \cos\theta, \quad b = -\sin\theta, \quad c = 0, \quad d = \sin\theta, \quad e = \cos\theta, \quad f = 0.$$

A translation by a vector with components (c, f) is an affine transformation with $a = e = 1, b = d = 0$. The rotations and translations of a plane also form a group, called the Euclidean group. This group is smaller than the whole group of affine transformations. Any element of the Euclidean group is a composition of a rotation and a translation. Thus it is described by three parameters: the angle of rotation and the two components of a translation. A general affine transformation **Aff** depends on six parameters, and it has fewer invariants (the 'things' left unchanged by the transformations) than the Euclidean group. In general, the bigger the group G, the fewer invariants it has. Every invariant of the affine group is an invariant of the Euclidean group, but the converse is, as we have seen, not true. Is there any advantage in taking this seemingly complicated approach to the classical Euclidean geometry? By starting from the Euclidean group, we can recover distances and angles as its invariants, even if we did not know what they were in the first place. This approach, which determines a geometry from its allowed group of transformation, has proved very influential in many areas of mathematics and goes beyond geometry.

Projective geometry

We devoted the whole of Chapter 5 to the projective geometry of perspective. Could this discussion have been shortened by putting the projective geometry in the framework of Klein's Erlangen programme? As we have seen, the key is to identify a group of allowed transformations. While we are aiming for a projective geometry of the plane, the starting point will be the group of transformations of the 3-space $\mathbb{R}^3$ taking a point P with coordinates (X, Y, Z) to $\bar{P}$ with coordinates $(\bar{X}, \bar{Y}, \bar{Z})$ such that

$$\bar{X} = aX + bY + cZ, \quad \bar{Y} = dX + eY + fZ, \quad \bar{Z} = gX + hY + iZ.$$

Equivalently, using matrices $\bar{\mathbf{v}} = A\,\mathbf{v}$, where $\mathbf{v}$ is a vector with components (X, Y, Z) and A is a 3 by 3 matrix,

$$A = \begin{pmatrix} a & b & c \\ d & e & f \\ g & h & i \end{pmatrix}.$$

We, shall moreover, assume that rows of this matrix are independent, which means that, when regarded as vectors, they do not lie on a two-dimensional plane in $\mathbb{R}^3$. This class of transformations is called *linear*. It depends on nine parameters $(a, b, \ldots, i)$, and is less general than the group of all affine transformations of the three-space, as it does not contain any translations. In Chapter 5 we defined the projective plane $\mathbb{RP}^2$ as the space of lines through the origin in $\mathbb{R}^3$, and any such line is in turn defined by two ratios of the homogeneous coordinates. How are these ratios changed under our group of transformations? Let us assume for a moment that Z is not equal to zero, and define two ratios,

$$x = \frac{X}{Z}, \quad y = \frac{Y}{Z},$$

so that, setting $\bar{x} = \bar{X}/\bar{Z}, \bar{y} = \bar{Y}/\bar{Z}$, we find

$$\bar{x} = \frac{ax + by + c}{gx + hy + i}, \quad \bar{y} = \frac{dx + ey + f}{gx + hy + i}.$$

This group of transformations is called the projective group. It transforms points on a plane, and is bigger than the group of affine transformations—a general projective transformation depends on eight parameters: it is not nine, as we have 'lost' one parameter by taking ratios. For example, if i is not zero, then we can divide all the parameters of a projective transformation by i without changing $\bar{x}$ and $\bar{y}$. This has an effect of setting i equal to one, and keeping the other eight essential parameters in the transformation. A seemingly more serious problem immediately arises: while we have assumed that Z is not zero, we have not said anything about $\bar{Z}$. Therefore, the denominator in the expressions for $\bar{x}$ and $\bar{y}$ may vanish, and then a point with coordinates (x, y) on the plane is transformed to infinity. However, this is exactly what we expect to see in projective geometry—the projective plane $\mathbb{RP}^2$ consists of an ordinary plane $\mathbb{R}^2$, together with a line at infinity. By starting off with a group of projective transformations, we have rediscovered the need for this line at infinity: the space S on which the projective transformations act has to be a projective plane, or the transformations would make no sense. The Klein perspective deduced the projective plane purely from the set of transformations. The projective group of transformations contains the affine group. Therefore, every invariant of the projective group is also an invariant of the affine group, but not vice versa; affine

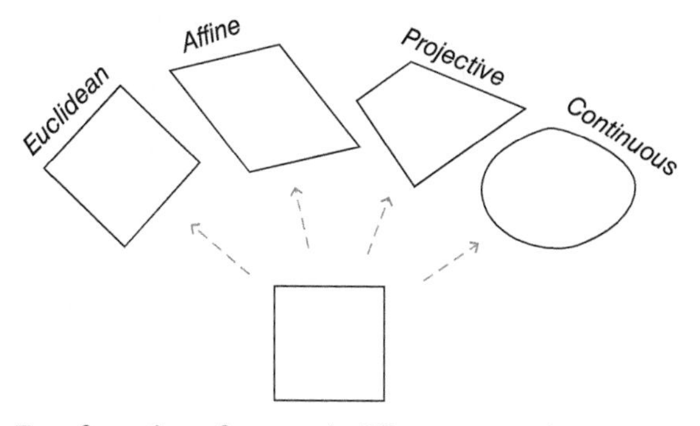

65. Transformations of a square in different geometries.

geometry is more restrictive than projective geometry. We have already shown that Euclidean geometry is in turn more restrictive than the affine geometry. This point was emphasized by Klein, who recognized projective geometry as the most general of the three, from the perspective of the transformation groups. We can also illustrate what happens to a square with side one under Euclidean, affine, projective, and general continuous transformations using a diagram (Figure 65).

Let us examine projective geometry in the simpler setup of a *projective line*. Our starting point is a linear transformation of a plane with coordinates $[X, Y]$. We define a projective line $\mathbb{RP}^1$ to be a set of lines through the origin in the plane. Any such line is determined by a ratio $x = X/Y$, where we have assumed that $Y \neq 0$. These ratios inherit their transformation rules from those on a plane. The relevant group consists of transformations of the form

$$\mathbf{M} \qquad \bar{x} = \frac{ax+b}{cx+d},$$

where (a, b, c, d) are parameters such that $ad - bc \neq 0$. These transformations are called Möbius, after the 19th-century German mathematician August Möbius. We define $\bar{x}$ to be ∞ if $x = -d/c$. So the arena for the projective transformations is $S = \mathbb{R} + \{\infty\}$—a line with a point at infinity added to it. This is exactly the projective line from Chapter 5.

What is it that we study in projective geometry of a projective line? In other words, are there any invariants of the group of projective transformations **M**? It turns out that there is only one such invariant. Any point on a projective line is uniquely described by its coordinate x, which we allow to be infinite. To any four points with coordinates (x_1, x_2, x_3, x_4), we can associate with their cross-ratio defined by

$$[x_1, x_2, x_3, x_4] = \frac{(x_1 - x_3)(x_2 - x_4)}{(x_1 - x_4)(x_2 - x_3)}.$$

It is true, and may be checked by an explicit calculation, that although under projective transformations all four points change according to **M**, their cross-ratio stays the same:

$$[\overline{x_1}, \overline{x_2}, \overline{x_3}, \overline{x_4}] = [x_1, x_2, x_3, x_4].$$

Another way to phrase this is as follows: any three points on a projective line can be mapped to any other three points by some transformation **M**. Therefore, we can choose a projective transformation which takes (x_1, x_2, x_3) to $(\infty, 0, 1)$. This requirement fixes the projective transformation uniquely, and the point x_4 is mapped to a point whose coordinate is the cross-ratio $[x_1, x_2, x_3, x_4]$.

Given two quadruples of points, they will be related by a projective transformation if and only if they have the same cross-ratios (Figure 66).

Topology

The biggest group of transformations of a plane consists of all *homeomorphisms*: these are continuous maps like stretching, bending, and twisting (but not tearing or gluing) which do not necessarily preserve distances or parallelism. So a cube and a sphere are homeomorphic in three-dimensional space, but a

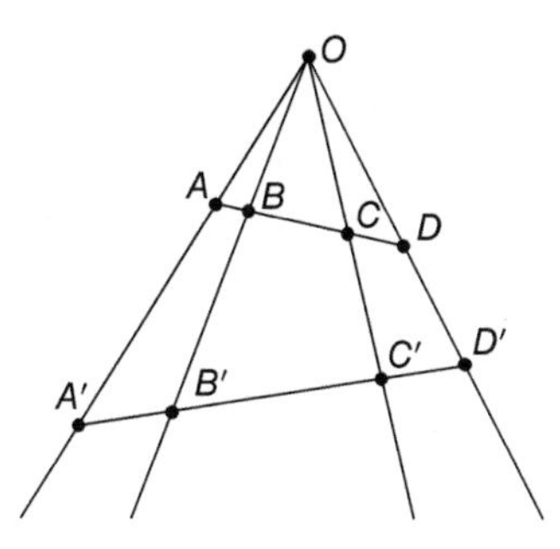

66. Two quadruples of points with the same cross-ratio are related by a projective transformation.

doughnut is not homeomorphic to either of them: it has a hole which cannot be got rid of by a continuous transformation. The corresponding Klein geometry is called *topology*. It has so few invariants that it evolved into a separate field of mathematics, and is not now regarded as a branch of geometry. It is therefore remarkable that some recent deep results in topology have been proved using techniques from geometry. The most prominent example may be Grigori Perelman's proof of the Poincaré conjecture. The conjecture, which was formulated in the early 20th century by the French mathematician Henri Poincaré, very roughly states that if S is three-dimensional space which is finite in size, consists of one piece, has no boundary, and has an additional property that all closed loops in this space can be continuously deformed to a point, then S can be continuously deformed into the sphere $x^2 + y^2 + z^2 + w^2 = 1$ in $\mathbb{R}^4$. Perelman's idea, which built on earlier work by Richard Hamilton, was to assume that the space S carries a notion of a curved Riemannian metric of the sort we discussed in Chapter 4. Once S is continuously deformed to another space, the metric also undergoes a deformation. By applying a tour de force analysis of the differential equation—the so-called Ricci flow—governing this metric deformation, Perelman was able to show that the metric evolves towards a metric of a constant Gaussian curvature, which can only exist on a sphere. He made his results available online in 2002 and 2003, but has never actually submitted them to a peer-reviewed journal. He has since withdrawn from mathematics, and refused many honours and awards offered to him—including the Fields Medal and a $1 million Clay Millennium Prize. In Perelman's own words:

> *Everybody understood that if the proof is correct, then no other recognition is needed.*

Geometry of Fermat's Last Theorem

One of the most famous problems in mathematics, which remained unsolved until relatively recently, was to determine whether there exist three natural numbers X, Y, and Z such that

$$X^n + Y^n = Z^n,$$

where n is a natural number greater than 2. The problem is known as Fermat's Last Theorem, and was named after the 17th-century French mathematician Pierre de Fermat, who wrote it in the margin of a book and stated that 'the margin was not big enough to include the marvellous proof' that no such natural numbers exist. The problem remained unsolved for over 300 years, until the British mathematician Sir Andrew Wiles presented a proof in 1995. The techniques used by Wiles belong to an area of *algebraic geometry*. While this book is not big enough to include his proof, I shall discuss the occurrence of geometry in the context of number theory.

If we divide both sides of Fermat's equation by Z^n, then the problem reduces to the question of whether there exist rational numbers $x = X/Z$ and $y = Y/Z$ such that the equation

$$x^n + y^n = 1$$

holds. This equation, when considered in Cartesian coordinates (x, y), gives a geometric shape known as a Fermat curve (Figure 67).

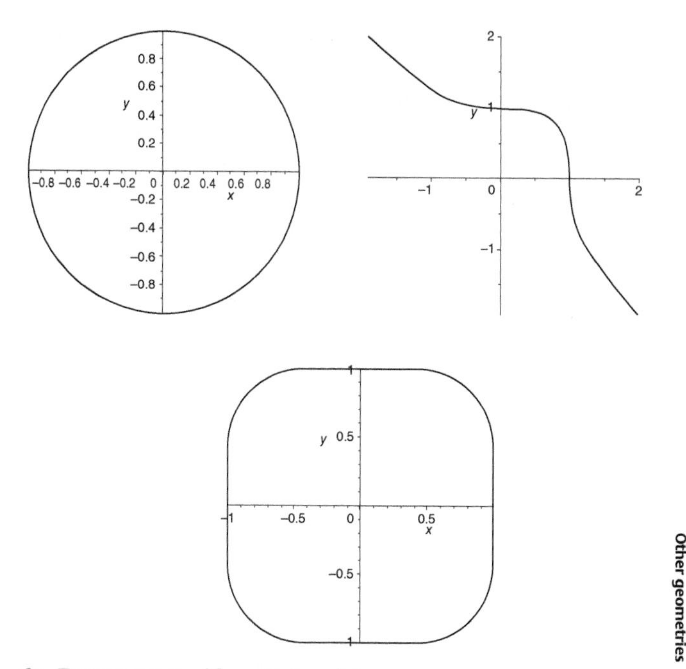

67. Fermat curves with $n = 2$, 3, and 4.

To appreciate the role of geometry in all this, let us first focus on the Fermat curve with $n = 2$, which is just a circle. Moreover, we know that integers X, Y, Z can be found in this case, as, for example,

$$3^2 + 4^2 = 5^2.$$

Triples of integers such as $(3, 4, 5)$ are called Pythagorean, as they appear as side lengths of a right-angled triangle. There are infinitely many such Pythagorean triples, so that there must be infinitely many points on the circle $x^2 + y^2 = 1$ such that the coordinates (x, y) are rational numbers. Such points are called *rational*. Let us see how they can be constructed by a geometric argument. We shall use a mechanical analogy, and regard the circle as a trajectory of a point moving on a plane, where the Cartesian coordinates x, y depend on 'time' t. Such a representation of a curve is called a parametrization. The parametrization we are seeking is rational, meaning that the functions x and y are ratios of polynomials in t. To obtain a rational parametrization, we shall identify points on the circle with their intersection with a straight line l passing through a fixed point on the circle (Figure 68). We shall choose this line to pass through the point $(-1, 0)$, so that

$$x^2 + y^2 = 1, \quad y = t(x + 1),$$

where in the second equation the gradient of the line l is given by a parameter t. To look for the points of intersection, substitute the second equation into the first:

$$x^2 + t^2(x + 1)^2 = 1.$$

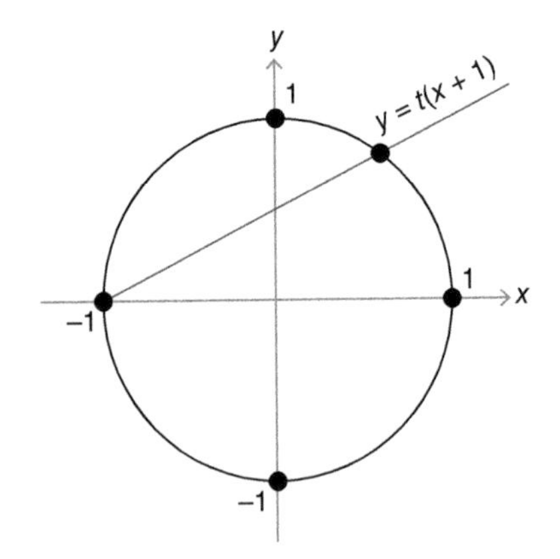

68. Rational parametrization of a circle.

There are two solutions for x: the first one, $x = -1$, corresponds to point $(-1, 0)$. The second one gives

$$x = \frac{1-t^2}{1+t^2} \quad y = \frac{2t}{1+t^2},$$

where the expression for y results from substituting x back into the equation of the line. The circle equation holds for all values of t because

$$x^2 + y^2 = \frac{4t^2 + (1-t^2)^2}{(1+t^2)^2} = 1.$$

Let us now go back to the Pythagorean triples. If both x and y are rational numbers, then so is $t = y/(x+1)$. Conversely, if $t = p/q$, where p, q are integers, then

$$x = \frac{p^2 - q^2}{p^2 + q^2}, \quad y = \frac{2pq}{p^2 + q^2}$$

are rational. As x and y have the same denominator, we can recover the Pythagorean triple as

$$X = p^2 - q^2, \quad Y = 2pq, \quad Z = p^2 + q^2.$$

Substituting integer values of (p, q) into these expressions gives infinitely many triples, the first few of which are

$$(3, 4, 5), \quad (5, 12, 13), \quad (8, 15, 17), \quad (7, 24, 25), \quad (20, 21, 29), \quad \ldots$$

It is remarkable that (apart from a few mistakes) a long list of such triples appears on a Babylonian clay tablet known as *Plimpton 322*, dated around 1800 BC—1,200 years before Pythagoras of Samos. It is not clear how the Babylonians found their triples, nor what they used them for.

This construction suggests a way of disproving Fermat's Last Theorem. If we find rational parametrizations of Fermat curves with $n > 2$, then (by substituting integer values of t) the integers X, Y, Z could be found which satisfy Fermat's equation. Given that we now *know* Fermat's Last Theorem to be true, something must be wrong with this argument. And indeed it is. First, it is not 'too hard' to show (at the level of an undergraduate course in algebraic geometry) that these Fermat curves do not admit a rational parametrization. If we tried to find such a parametrization by intersecting the Fermat curve with a line through a fixed point on this curve, the method above would lead to a certain polynomial equation of degree n. The procedure of finding the roots of this equation cannot be accomplished by factorization. In 1983 the German mathematician Gerd Faltings proved that curves given (like the Fermat curve) by vanishing of a polynomial in x and y of degree n have at most finitely many rational points if $n > 3$. This implies that there may only be finitely many integers (X, Y, Z) satisfying the Fermat relation if $n > 3$. This was the closest anyone came to proving Fermat's Last Theorem, until in 1995, when Andrew Wiles showed that the number of such integers is in fact zero. Faltings was awarded the Fields Medal—one of the highest prizes a mathematician can receive—for the proof of several theorems, including the one stated above. Fields Medals are awarded every four years to up to four mathematicians under the age of 40. Andrew Wiles was 42 years old at the time when the correct version of his proof was announced, and therefore missed out on a medal.

A careful reader would have noticed that the case when $n = 3$ appears to be special. The existence of infinitely many rational points is not excluded by the Faltings result. The statement of Fermat's Last Theorem for $n = 3$ pre-dates Fermat by over 600 years. It was noted by the 10th-century Persian astronomer Al-Khujandi. Al-Khujandi's attempted proof of the theorem was incorrect. A correct proof was established by Leonhard Euler in 1760.

The example of the Fermat curves suggests an exciting possibility that algebraic geometry may be of service to mathematics and offer ways of solving long-standing problems with its techniques. A slight drawback is that in the second half of the 20th century algebraic geometers have, in search of rigour, developed a technical and hermetic language. They now have a reputation among other mathematicians similar to that which mathematicians have among non-mathematicians. This, in turn, was well summarized by Johann Wolfgang von Goethe:

> *Mathematicians are like Frenchmen: whatever you say to them they translate into their own language and forthwith it is something entirely different.*

The Polish-Jewish mathematician Hugo Steinhaus included the following problem in his collection of recreational mathematics problems aimed at non-specialists:

> **Problem of Steinhaus.** Does there exist a square with integer sides and a point P in the plane of this square such that the four distances from P to the vertices of the square are integers?

Steinhaus included solutions to all problems from his collection except the one above, which he admitted he did not know how to solve. By introducing a Cartesian coordinate system the reader will, with little difficulty, translate this problem into the statement about the existence of rational solutions to a number of quadratic equations. It is not known whether any such solutions exist.

Chapter 7
Geometry of the physical world

For over two millennia, Euclidean geometry provided a framework for interpreting empirical observations and making accurate predictions, for phenomena like solar eclipses. However, as our measuring apparatus improved, it became clear that the geometry of the Universe is not that of Euclidean space. Twentieth-century theories of physics, including Einstein's general and special relativity, are based on geometries different to Euclidean geometry that describe the Universe more accurately. The most basic of these is Minkowski geometry, which replaces Pythagorean distance by a space-time interval, where the squared interval s between two points (called events in this chapter) P and P' with space-time coordinates (t, x, y, z) and (t', x', y', z') is

$$s(P, P')^2 = c^2(t - t')^2 - (x - x')^2 - (y - y')^2 - (z - z')^2,$$

where the constant $c = 299,792,458$ metres per second is the speed of light.

The theory of gravitation is based on a curved version, in the sense described in Chapter 4, of the Minkowski distance. It turns out that electromagnetic, as well as weak and strong nuclear, interactions are also manifestations of curvature, but this time in the internal spaces of charges (in the case of electromagnetism) or so-called *colours* in the case of the quantum theory of strong interactions holding quarks together. It may well be that there is some kind of geometry which unifies the geometry of Einstein's gravity with those of the other three fundamental interactions, but at the time of writing we have no idea what this geometry would look like. In this chapter I shall therefore only explore the geometries (rather than the underlying physical theories, which are the subjects of other *Very Short Introductions*) of general and special relativity.

Special relativity

The foundations of special relativity were laid down by Albert Einstein in 1905, but it was the German mathematician Hermann Minkowski who recognized that the geometrical framework of Einstein's theory is a four-dimensional space-time. We shall call this space-time $\mathcal{M}$. The points of $\mathcal{M}$ are called events. Events are idealized building blocks of the fabric of the Universe. To specify an event, one gives its location in space as well as the time of its occurrence. This information is encoded in four Cartesian coordinates (t, x, y, z).

One of the properties of distance in Euclidean, as well as the spherical, hyperbolic, and Riemannian geometries that we encountered in the earlier chapters, is positivity. The distance between two points is a nonnegative number which can be zero if and only if the two points coincide. The distance in Minkowski geometry does not have this property. For example, the space-time interval between two points P and Q with space-time coordinates $(1, 0, 0, 0)$ and $(0, c, 0, 0)$ vanishes as $s(P, Q)^2 = c^2(1-0)^2 - (0-c)^2 = 0$.

Let P be any point in the Minkowski space. The interval function divides the Minkowski space into three regions, R_+, R_0, and R_- (Figure 69), which consist of all points Q such that $s(P, Q)^2 > 0$, $s(P, Q)^2 = 0$, and $s(P, Q)^2 < 0$, respectively. The horizontal plane on Figure 69 represents the spatial directions with coordinates (x, y, z). This plane is three-dimensional, but appears as

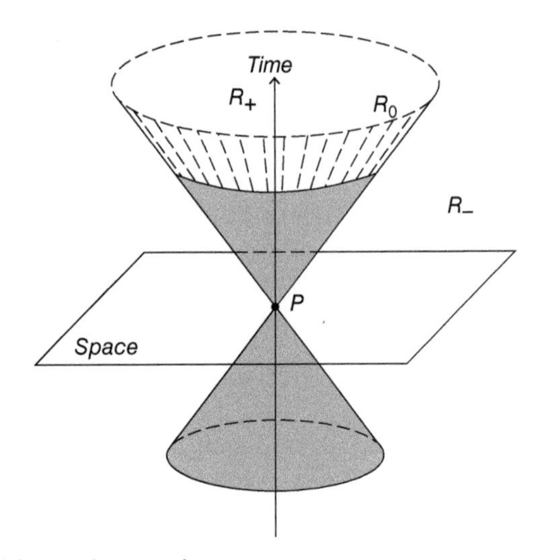

69. Light cone in space-time.

two-dimensional on the picture. Region R_0 is the three-dimensional surface of a cone, called the *light cone* of P, in four-dimensional space-time. To see that it is indeed a cone, assume that the reference frame has been chosen so that the coordinates of P are $(0, 0, 0, 0)$, and the coordinates of Q are (t, x, y, z). Then the condition $s(P, Q)^2 = 0$ is equivalent to

$$c^2t^2 - x^2 - y^2 - z^2 = 0,$$

which is the equation of a cone in four-dimensional Cartesian coordinates.

The region R_0 is further divided into the future light cone, where $t > 0$, and the past light cone, where $t < 0$. This terminology is motivated by special relativity, where the light cone at P consists of light rays passing through P. According to quantum theory, light consists of massless particles called photons. Thus a photon emitted at P will travel along the future light cone, and a photon absorbed at P reached it along the past light cone. All points inside the light cone belong to R_+. In special relativity nothing, including information, can travel faster than light. Therefore, only events in regions R_+ and R_0 with $t < 0$ could have influenced P. Similarly, a signal sent from P can only reach R_+ and R_0 with $t > 0$. Moreover, R_0 can only be reached by a photon emitted at P, and a massive object—referred to as an observer—is confined to R_+. The *causal future* of P consists of R_+ and R_0 with $t > 0$. Events outside the causal future of P could not have been influenced by P in any way. Similarly, the *causal past* of P has $t < 0$. Events in R_- are not causally related to P. No information about this region will ever reach P, and no signal sent from P can reach R_-.

The future light cone is the boundary region of the causal future of P. The slope of this light cone is given by the speed of light c, and it is customary to align it at 45°. A trajectory of a massive particle moving with constant velocity of magnitude $v < c$ corresponds to a straight line through P and confined to R_+. To sum up, the geometry of the four-dimensional Minkowski space is richer than that of the four-dimensional Euclidean space $\mathbb{R}^4$. While both spaces are endowed with notions of distance, the Minkowski space has the causal structure which is absent in the Euclidean geometry.

Recall that a straight-line segment in Euclidean space minimizes the distance between its two endpoints. Consider instead extremizing the interval between two events P and Q, such that Q is inside the light cone of P (Figure 70). The straight-line segment inside R_+ joining P to Q leads to the positive interval $s(P, Q)^2$. If we instead join P and Q with line segments lying on the light cones and add the corresponding intervals (dashed lines in Figure 69), then the resulting interval is zero. Moreover, joining P and Q zigzagging along line segments in R_- we can make the interval $s(P, Q)^2$ as negative as we want to! Therefore, in Minkowski geometry a straight-line segment joining P and another event Q in the region R_+ maximizes the space-time interval between P and Q. This is a consequence of the reversed triangle inequality, where the

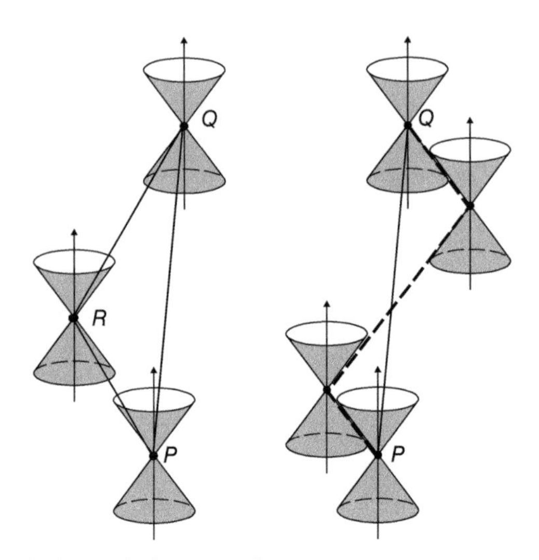

70. Geodesics are the longest paths.

segments of a triangle lie inside the light-cones of their vertices (Figure 70).

$$s(P, R) + s(R, Q) \leq s(P, Q).$$

Curves in space-time are called world lines. Thus a world line is different from a trajectory in space, as it also incorporates the time direction. World lines of photons are confined to light cones, and world lines of massive observers (for example astronauts, but also elementary particles) moving with constant velocity and passing through an event P are straight lines in the region R_+.

One consequence of the geometry considered thus far is that time-simultaneity is a notion relative to an observer; there is no absolute notion of time in special relativity. To see this, consider two observers O and O'—call them Alice and Bob—who meet at P and agree that the time coordinate at P is zero. Let P_1 and P_2 be

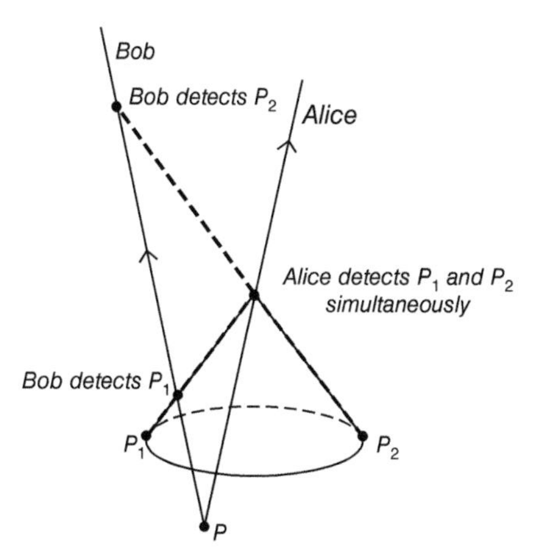

71. Simultaneity is not absolute.

two events which are simultaneous according to Alice. This means that the word lines of photons emitted at P_1 and P_2 intersect the world line of Alice at the same point. The second observer, Bob, will detect the photon emitted at P_1 before he detects the photon emitted at P_2 (Figure 71). He will reckon that P_1 happened before P_2. We could colour this picture by adding a third observer who, by similar reasoning, will perceive P_2 before he sees P_1. The three observers will fundamentally disagree about the time ordering of events P_1 and P_2, and yet their calculation of the space-time interval $s(P_1, P_2)^2$ will give the same answer.

The three-dimensional planes of simultaneity differ for Alice and Bob, but they are related by a transformation which I shall now discuss. Let us recall the Cartesian coordinate approach to Euclidean geometry, and assume that we want to describe a tetrahedron in three-dimensional space but by using two coordinate systems with the same origin. The Cartesian coordinates of the vertices of the tetrahedron will be different in both coordinate systems, but the distances between the vertices will not depend on the choices of coordinates. Both coordinate systems are related by a rotation which is an isometry of space. If the form of this rotation is given (for example as a 3 by 3 matrix), then the coordinates of the tetrahedron in one system can be found by applying a transformation to the coordinates in the second system. The isometries of the Minkowski space which preserve the space-time distance consist of four translations in the x, y, z, and t directions, three spatial rotations of the (x, y, z) space, and three transformations called the Lorentz boosts, which mix up the time and space coordinates O and O' used by Alice and Bob. If Bob travels with velocity v in the x direction relative to Alice, then the Lorentz boost between the two coordinate systems O and O' is

$$t' = \gamma\left(t - \frac{vx}{c^2}\right), \quad x' = \gamma(x - vt), \quad y' = y, \quad z' = z, \quad \text{where } \gamma = \frac{1}{\sqrt{1 - \frac{v^2}{c^2}}}.$$

These transformation rules imply that if an object moves at the speed of light in the coordinate system O, then it moves with the same speed in the coordinate system O'. The speed of light is the same for all observers. Another consequence of the Lorentz boost is a time dilation. If a clock used by Alice ticks at time intervals T, then the tick events take place at points with coordinates $(cT, 0, 0, 0), (2cT, 0, 0, 0), (3cT, 0, 0, 0)$, and so on. In the coordinate system O' used by Bob, the interval between the ticks is $T\gamma$, which is longer than in the frame of Alice moving with zero velocity.

Similarly, the spatial directions undergo contractions when one passes between the coordinate systems of Alice and Bob. A tetrahedron on the $t = 0$ plane relative to O corresponds, after a Lorentz transformation, to another tetrahedron but with smaller spatial area and distorted angles in the coordinate system O' used by Bob. In fact, the time contraction together with the absolute value of the speed of light implies that space contraction must also take place. This can be illustrated as follows: if Alice, who inhabits planet Earth, uses a second as a unit of time, then the *light second*, equal to 299,792,458 metres, is her natural unit of distance. This is the distance travelled by light at the absolute speed c in one second. In these units, the Sun is around eight light minutes from the Earth. For all Alice knows, the Sun may have exploded a minute ago, but she would not have noticed anything unusual, as the light did not have enough time to reach her. However, as we have seen, the second measured by Alice was different to that measured by Bob. Thus the meaning of the distance eight light minutes is also observer-dependent.

There is another, geometric approach to Lorentz transformations which brings up the projective geometry from Chapter 5. Think of Alice looking at the stars in the night sky. The points she sees in the sky are intersections of the world lines of stars with Alice's own past light cone. These are points on the celestial sphere which arise as the intersection of the light cone $c^2t^2 - x^2 - y^2 - z^2 = 0$ with the plane $t = -t_0$, where t_0 is the time it took for the light to travel between the stars and Alice. If we take t_0 to be -1, then this intersection is indeed a sphere in $\mathbb{R}^3$ of radius c. Different choices of t_0 merely change this radius. Let Bob be another observer moving with a constant velocity v with respect to Alice. If the world lines of Alice and Bob intersect at a point P, then both observers at P will perceive the same stars but located at different points on their respective celestial spheres. The Lorentz transformation which takes the celestial sphere of Alice to that of Bob is

$$\zeta' = \frac{a\zeta + b}{c\zeta + d},$$

where ζ is the complex coordinate obtained by the stereographic projection from the sphere $x^2 + y^2 + z^2 = c^2$ to the $z = 0$ plane (recall Figure 48 from Chapter 4). This is nothing but a Möbius transformation, which we encountered in Chapter 6, but now mapping a complex projective line to another such line. The complex numbers a, b, c, d are the parameters of the Lorentz transformation, and are subject to one relation, $ad - bc = 1$. Therefore, only three out of the four complex parameters can be specified arbitrarily. The transformation depends on six real parameters, two for each of the complex numbers. Three of them are the angles of spatial rotations on three-dimensional planes of simultaneity, and the other three correspond to Lorentz boosts.

Geometry of Newtonian space-times

Let us compare the geometric structures of Minkowski space-time with those arising in the theories of space and time pre-dating special relativity. Contrary to a popular belief, these pre-relativistic space-times are also four-dimensional, but they have an absolute, observer-independent notion of time.

The geometry resulting from an assumption that both space and time are absolute is called the Aristotelian space-time, $\mathcal{A}$. In Aristotelian geometry an observer at rest is preferred to all other observers moving with constant velocity. This has presumably resulted from the Greek belief that the Earth occupies a special place in the cosmos. Therefore, $\mathcal{A} = \text{Time} \times \text{Space} = \mathbb{R} \times \mathbb{R}^3$. The Aristotelian space-time distance between Alice at midday and Bob at 6 p.m. is given by two numbers: the time interval between the two events equal to six hours, and the Euclidean distance between the two observers as the crow flies.

The next geometry in the ladder is the Galilean space-time $\mathcal{G}$ used in the Newtonian theory of gravitation. The laws of physics in the Galilean space-time can also be formulated in geometric terms. This has been clarified by the French mathematician Élie Cartan. Time is absolute in $\mathcal{G}$, but space is not, in that one cannot make sense of distances between points unless they lie on the same plane of simultaneity. The Aristotelian way of computing the distance between Alice and Bob does not make sense in $\mathcal{G}$—we can only say how far apart Alice and Bob are at any given time (Figure 72).

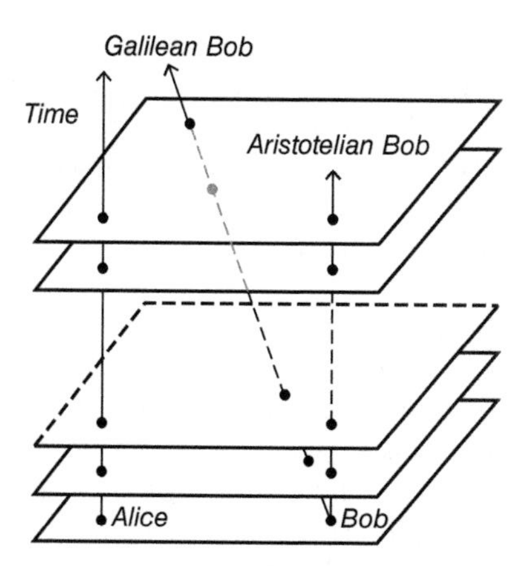

72. Distance in Aristotelian and Galilean space-times.

The notion of space is different for each instant of time, but there exists a projection of $\mathcal{G}$ to the time axis $\mathbb{R}$ associating the time coordinate to any point in the Galilean space-time. Such geometric structures are called fibre bundles. The fibre of each point of the time axis $\mathbb{R}$ is a copy of three-dimensional Euclidean space $\mathbb{R}^3$, together with its Euclidean distance.

In the Minkowski space-time of special relativity, the speed of light and space-time intervals are both absolute, but neither space nor time are.

The space-time of general relativity

The geometric formulation of special relativity was built on observational data suggesting a need for a new theory of space and time. By contrast, Einstein's general relativity—the theory of gravitation consistent with special relativity—was driven not by experiments or observations, but to a large extent by geometric aesthetics. Einstein's great insight was that gravitational attraction is an effect of space-time curvature. The bending of a comet trajectory by the Sun's gravity can be explained by postulating that the comet is in fact moving along a geodesic which maximizes the space-time interval but in the presence of curvature. The resulting path will appear to us as an ellipse. It is important to stress that this ellipse is a geodesic not in a three-dimensional space, but in four-dimensional space-time.

The theory emerged in 1915, but it took several decades to come to terms with its mathematical complexities, some of which have still not been resolved. The geometric passage from the Minkowski space-time to the Einsteinian space-time of general relativity is analogous to the step from Euclidean geometry to a curved geometry. In Einstein's theory of gravity, the Universe is a four-dimensional manifold $\mathcal{E}$, and it may not be possible to cover it with a single coordinate system. The Minkowski distance is replaced by a curved metric g of a sort we discussed in Chapter 4, but with the proviso that the distance function resulting from this metric does not have to be positive definite. Such metrics are called pseudo-Riemannian. A new feature resulting from dropping the positivity of the distance is the possibility that two non-identical points in space-time can be connected by a curve of zero length. Such curves are called null. Light travels throughout the space-time along geodesics, which are also null curves.

As in Minkowski space, the metric of general relativity equips space-time with a causal structure given by light cones, but more assumptions need to be made on the pair $(\mathcal{E}, g)$ to rule out the pathologies associated with time travel. We did not have to worry about it in the Minkowski space, where the world line of an observer cannot start and end at the same point in Figure 69, as otherwise it would have to cross from R_+ to Region R_-, which is ruled out, as signals and massive particles cannot travel faster than light. The situation is more subtle if the light cones are deformed by the presence of curvature and the space-time $\mathcal{E}$ is allowed to be a manifold. Imagine $\mathcal{E}$ as a four-dimensional doughnut, where the light-cones go around the hole in the middle and a time-like path could be taken by an observer to return to its starting point. This leads to a plethora of paradoxes. Alice moving along such a curve could eventually reach her own past, thus opening possibilities to interesting scenarios, some of which have been explored in the *Back to the Future* movie trilogy. It is assumed that none of this can take place.

In general relativity, no particular coordinate system is preferred, and therefore the laws of physics with their geometric description need to be formulated in a coordinate-free way. This is achieved by employing tensor calculus. A vector, a pseudo-Riemannian metric, and its curvature are all examples of tensors. A tensor is a collection of numbers (for example components of a vector in some coordinate system) together with a rule on how these numbers transform if one changes from one coordinate system to another. This rule is such that if a tensor vanishes in one coordinate system, it vanishes in all systems. Thus the property for space-time to be curved at some point is coordinate-independent. The Riemann curvature introduced in Chapter 4 naturally splits into two parts, called Ricci and Weyl, and satisfies the Einstein equations:

$$\begin{pmatrix} \textbf{Einstein curvature} \\ \textbf{of space-time} \end{pmatrix} = \begin{pmatrix} \textbf{Matter and pressure} \\ \textbf{density of the Universe} \end{pmatrix}.$$

Skipping over details, the left-hand side of the Einstein equations is a tensor—called the Einstein tensor and made out of, but not exactly equal to, the Ricci curvature—so the right-hand side has to be one too. It is called the energy–momentum tensor. This tensor may also contain the cosmological term causing the Ricci curvature to increase with distance. Mass concentrated in some region of space-time contributes to the Ricci curvature, which in turn has an effect of reducing the volume of this region. The American physicist John Archibald Wheeler summarized these equations well in one sentence:

Matter tells space-time how to curve and curved space-time tells matter how to move.

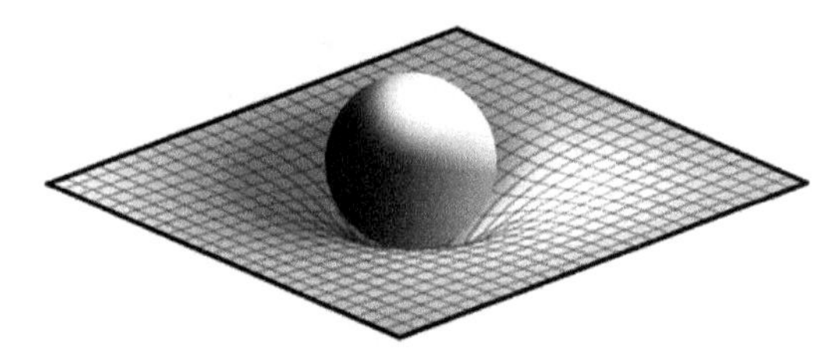

73. Matter curving space-time.

A star moving in a galaxy will curve the space-time around it. This curvature will in effect modify geodesics, and thus the trajectory of the star (Figure 73).

The Weyl curvature is not determined by the Einstein equations, and corresponds to gravitational degrees of freedom. In particular, a metric may satisfy the Einstein equations in a space-time with no matter content, and nevertheless be curved because of the presence of Weyl curvature. The Weyl curvature distorts the geometry of space-time in a way analogous to how the Moon causes tides in the ocean. It exerts tidal forces on particles and light moving through the space-time. This effect preserves the volumes of space-time regions affected by the curvature, but changes their shape.

The unknown in the Einstein equations is the space-time metric. As the Einstein curvature depends on derivatives of the metric, the Einstein equations fall into the class of differential equations and form one of the trickiest systems of such equations known to science. Not only is the metric unknown, but also the underlying manifold $\mathcal{E}$ supporting this metric needs to be found. The equations are non-linear, which is a technical term signalling to students and researchers that all methods they have acquired at school and university to solve differential equations are likely to fail. One solution to the Einstein equations is Minkowski space. The curvature vanishes for this solution, and so there is no matter content in the Minkowski Universe. Thus, as was to be expected, the special relativity falls into the framework of general relativity. Very few other solutions have been found—each such solution does in principle give a candidate for a Universe, and we shall shortly discuss one in some detail—so the research has focused on finding the qualitative properties of solutions. The idea is to specify the initial data consisting of the space-time metric and its rate of change on some three-dimensional surface of constant time, and see whether the Einstein equations fully determine the evolution of such initial data. If they did, then general relativity would be a deterministic theory, with Einstein equations governing the dynamics of space-times: you tell me the data at time equal to zero, and I (with the help of my research students, postdocs, and supercomputers) will in principle be able to solve the Einstein equations to tell you what this metric will be in the future. Unfortunately, we now know of situations where this predictability of general relativity does not hold. There exist solutions of Einstein equations called black holes, where general relativity predicts its own failure. We shall discuss them next.

Geometry outside a black hole

As in special relativity, the space-time $\mathcal{E}$ of general relativity can be thought of as a collection of light cones, one cone through each point in $\mathcal{E}$. The cones are easy to visualize—start at an event *P* and draw the locus of all events with zero space-time interval from *P*. Therefore, knowing the space-time intervals between all space-time events definitely determines the cones. On the other hand, the cones do not completely determine the interval, but they will suffice for what I want to explain. Only from the knowledge of the cones can we determine whether *P* and *Q* have zero interval between them: this happens if and only if *P* belongs to the light cone of *Q*, or *Q* belongs to the light cone of *P*. Similarly, *P* and *Q* can be connected by a geodesic of a positive length (such geodesics are called time-like) if *P* lies inside the cone of *Q*, or *Q* lies inside the cone of *P*.

While the light cones in Minkowski space have the same slope given by the speed of light, the cones in curved space-times can be bent by space-time curvature. Just a year after general relativity was discovered, while he was serving in the German army during the First World War, the German physicist Karl Schwarzschild found a solution to the Einstein equations that shows this (Figure 74).

The time direction goes upwards, and one of the three-space directions has been suppressed. Light cones far from the centre of the Figure look like those in Minkowski space, which has no curvature. Moving away from the flat region towards the three-dimensional cylindrical surface $\mathcal{H}$ called the event horizon, the curvature effects become relevant and the light cones bend towards $\mathcal{H}$. In the limiting case, when a point P lies on the horizon, the light cone of P is tangent to $\mathcal{H}$. All future-pointing time-like and null curves through P (with the exception of null curves

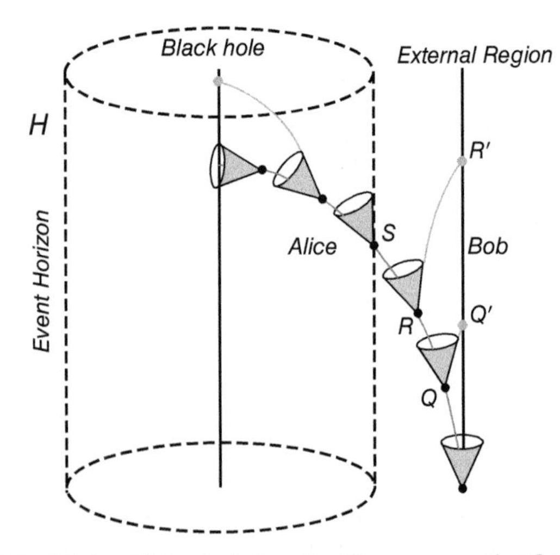

74. A black hole with the dashed vertical lines representing the event horizon.

tangent to $\mathcal{H}$) will enter the region bounded by $\mathcal{H}$. If a point P is inside the horizon, then future-pointing null or time-like geodesics through P cannot leave the region inside the horizon. Thus no information about the region inside the horizon can ever reach observers outside the horizon. The event horizon is not made of any matter. It is a three-dimensional surface which can only be crossed in one direction, and separates the region in space-time from which the light rays can escape to infinity, and the region from which no signal can escape. It is this region—bounded by the cylinder $\mathcal{H}$ in Figure 74—which is referred to as a black hole. The geodesics in the black hole space-time are different to those in the Minkowski space. This effect, which becomes stronger near the event horizon, can be illustrated by the following example. Alice and Bob meet at some event outside the horizon. Alice decides to venture into the black hole, and Bob will stay outside. Alice sends Bob updates from her journey by emitting light rays every second as measured by her. Geometrically, this comes down to dividing Alice's trajectory into segments with equal space-time intervals, and drawing a null geodesic away from the horizon from the endpoint of each segment. As Alice's light cones tilt near the horizon, so does the slope of each light ray she emits. Consider two signals emitted by Alice at events Q and R. Bob detects them at events Q' and R', but the space-time intervals satisfy

$$s(Q,R)^2 < s(Q',R')^2,$$

so that according to Bob, the frequency of the signals sent by Alice goes down as she approaches the horizon. The closer Alice gets to the horizon, the stronger this effect becomes. Alice finally crosses the horizon at the event S, and duly sends a message to Bob from there. He will, however, never receive it, as the light ray through S will stay tangent to the horizon. From Bob's perspective, Alice slows down as she approaches the horizon, without ever getting there. Bob will receive the message sent by Alice just before she reached S—he may have to wait a long time, perhaps millions of years, according to his clock (that would depend on the mass of the black hole, and on how far Bob and Alice were from the horizon when they parted)—but this will be the last message he will have received. Alice, on the other hand, will carry on with her journey towards the now-unavoidable singularity. The path along which Alice travels is at a finite space–time interval from the singularity, so Alice will reach the singularity in finite proper time (recall that the proper time is just the interval divided by the speed of light). Then space and time will end for Alice. Our current understanding of general relativity offers Alice no guidance as to what happens next.

The Schwarzschild solution to Einstein's equations describes a time-independent gravitational field of a spherically symmetric body of mass m. This body could be a star which collapsed under its own gravity. If the star is sufficiently massive, the gravitational force dominates the pressure of the matter inside the star, and the star collapses down to a singularity of infinite density at $r = 0$, where r is the radial coordinate centred in the middle of the cylinder in Figure 74. Thus a black hole is the final stage of gravitational collapse; the matter of a star collapses inwards through the surface of a three-dimensional curved cone (Figure 75). Once the spatial section of this cone crosses the Schwarzschild radius $r = 2m/c^2$, it becomes the event horizon and the star turns into a black hole. The geometry of space-time from this moment onwards is represented in Figure 74.

Putting m equal to the mass of the Earth into the expression for the Schwarzschild metric, we can compute that the radius of the event horizon is equal to about one centimetre. Thus the whole Earth would need to be squeezed into the volume of a small walnut to create a black hole. There is now direct observational evidence that black holes of masses equal to 10 or more solar masses exist in the Universe. Some black holes are far heavier than that. On 10 April 2019, the team of scientists behind the Event Horizon Telescope released an image of a black hole of mass exceeding six billion solar masses in the centre of the M87 galaxy.

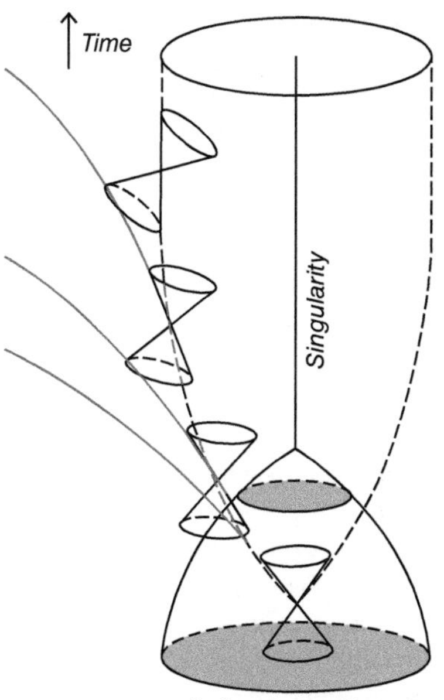

75. Gravitational collapse of a star to form a black hole.

The horizon of this enormous black hole is about 18 billion kilometres from its centre, and the gravitational force at this horizon is about a quarter of the Earth's gravity. An observer crossing such a horizon would not experience any unusual effects. Thus, for all you know, the Earth may be falling into a black hole as you are reading this. If that is indeed happening, you are doomed. In the next section I shall explain how the geometry of space-time ends inside a black hole.

Geometry inside a black hole

The expression for the Schwarzschild metric becomes invalid at the surface of the horizon. This apparently singular behaviour is also indicated by the vertical slope of the light cones tangent to the horizon. The gradient of these slopes is given by the formula

$$\pm\frac{r}{r - 2m/c^2},$$

which is close to one (and corresponds to 45°, which is familiar from special relativity) if r is large, and then increases as r becomes smaller, and finally blows up at $\mathcal{H}$, where $r = 2m/c^2$. The curvature of the Schwarzschild space-time is, however, well defined, and finite at $\mathcal{H}$. This initially caused a lot of confusion, which took over 40 years to clear! It is now understood that the metric can be made completely regular at the horizon if different coordinates are chosen. This is allowed, as, after all, a space-time is a manifold and more than one coordinate system may be needed to describe it.

The second problematic region for the Schwarzschild solution is the thick vertical line in Figure 74 corresponding to the zero radial distance $r = 0$. This is an intrinsic singularity where the Riemann curvature becomes infinite. It cannot be removed by passing to different coordinates, and a thorough analysis of what happens with geodesics near $r = 0$ gave rise to the geometric theory of singularities for other solutions to the Einstein equations. A problem with defining a singularity as a region in space-time where the curvature is infinite is that one may remove such a region from the manifold and declare space-time to be what is left after the surgery. The resulting space-time has finite curvature, but another pathology emerges: there exist geodesic curves which cannot be extended indefinitely. Such geodesics are called 'incomplete'. An illustration can be a surface of a cone in the Euclidean 3-space, with the vertex removed. The curvature of this space-time is everywhere finite, and in fact equal to zero, but the geodesics which are the generators of the cone have nowhere to go once they have reached the edge. Therefore, one cannot get rid of the singularity by simply removing the vertex of the cone.

It took mathematicians some time to come up with a good definition of what a singularity is. In 1965 Roger Penrose established a theorem which stated that, once several reasonable assumptions are made, the space-time of general relativity must contain a singularity. This theorem, for which Penrose was eventually awarded the Nobel Prize in Physics in 2020, was subsequently improved and extended by Penrose together with Stephen Hawking. The assumptions behind the Penrose–Hawking theorems are that there are no closed time-like curves, and that the gravitational force is attractive, and strong enough so that there exists a region of space-time such that all null and time-like geodesics are trapped in this region.

Thus the situation we have analysed in Figure 74 is not—as was initially thought—an artefact of spherical symmetry of the Schwarzschild solution. It is generic: there is a black hole singularity at the final stage of the gravitational collapse of massive stars, resulting in the end of time for particles moving along incomplete geodesics. Such particles reach, in finite time, the region where neither time nor space makes sense. This is the inevitable fate of any observer crossing the event horizon.

Another kind of singularity which arises in general relativity is the initial singularity present at the beginning of the Universe, called the Big Bang. The theorem explaining the initial singularity was proved by Stephen Hawking in his doctoral thesis by reversing the direction of time in Penrose's argument and turning the gravitational collapse into expansion of the Universe.

The Penrose–Hawking theorems show the inevitability of singularities, but offer no guidance as to their nature, except that the laws of physics as we know them, and their geometric description break down. A new type of geometry may be needed to describe the physics inside black holes, and some insight into this new geometry could perhaps be gained by observing what happens to particles as they reach the singularity. However, this is not possible for an observer outside the event horizon, as no signal from the singularity can pass through the horizon surface. Roger

Penrose has put forward the *cosmic censorship hypothesis*, which states that a gravitational collapse always results in a black hole: all singularities are hidden inside an event horizon. To paraphrase, there exists a cosmic censor preventing the creation of a naked singularity which can be observed directly. Despite its theological and philosophical appeal, this hypothesis is a highly technical problem in mathematical analysis and differential geometry. It remains one of the biggest unsolved problems in general relativity.

On the space-time continuum

The Minkowski and Euclidean geometries as well as their curved versions discussed in this book are based on concepts like a point (which has no dimension or size), or an infinitely thin line. Such notions belong in the Platonic world of mathematics; no constructions based on these ideal concepts are realized in Nature. Matter and energy as we currently understand them are quantized—they consist of discrete portions or quanta which cannot be divided into smaller portions. Thus the pencil in the compass needed for the geometric constructions of Euclid is never infinitely sharp, and the ruler is never exactly straight. For this and other reasons, the perfect shapes of geometry are only approximations of shapes in nature which, at some microscopic level, have a grain-like structure.

We have not worried about this, as although the presence of matter introduces discreteness, the underlying fabric of space-time is continuous and can be understood geometrically. It may be that, at some level, the very notion of a space-time point will have to be abandoned. There is no experimental evidence that we should do it just yet, but there exist some more or less radical proposals which give up the notion of points as fundamental objects. One such proposal is twistor theory, where space-time points arise as derived objects and light rays play a fundamental role. Twistor space is a six-dimensional manifold whose points correspond to light rays, or twisting families of light rays in the

Minkowski space. The points in Minkowski space correspond to two-dimensional spheres in the twistor space, which makes the correspondence non-local in a way similar to the projective duality we discussed in Chapter 5. The whole of special relativity, and some aspects of general relativity, can be reformulated in twistor space in a remarkable way which brings up complex numbers and puts classical and quantum physics on equal footings. It remains to be seen whether the twistor approach sheds light on the nature of space-time singularities, or indeed whether a final theory of space and time—if one is ever found—is geometric in its core.

Further reading

While few will read Euclid's *Elements* from beginning to end, it is regarded as the most influential mathematics book ever written. Most currently available editions (e.g. Dover, 2nd Edition Unabridged) are early-20th-century translations by Thomas Heath, and contain his interesting commentary. A lot of insight into subsequent developments can be gained from reading about the history of the subject in *A History of Geometrical Methods* by J. L. Coolidge (Dover, 2003).

Geometry and Imagination by D. Hilbert and S. Cohn-Vossen (2nd Edition, AMS Chelsea, 1999) is based on lectures given by Hilbert in 1921, and contains a lot of classic material and very few formulae. H. S. M. Coxeter's *Introduction to Geometry* (2nd Edition, Wiley Classics Library, 1989) is a proper university-level maths textbook with theorems and proofs, but accessible to readers with a high school level mathematics education. John Roe's *Elementary Geometry* (OUP, 1993) is aimed at a similar level.

In order to keep this book at elementary level, I have consistently avoided calculus. The only branch of geometry likely to have suffered from this omission is differential geometry presented in Chapter 4. Roe's book can fill this gap, and so can M. P. do Carmo's *Differential Geometry of Curves and Surfaces* (2nd edition, Dover, 2016).

Readers interested in the interplay between symmetry and geometry should consult *Symmetry: A Very Short Introduction* by I. Stewart (OUP, 2013) and *The Symmetries of Things* by J. H. Conway, H. Burgiel, and C. Goodman-Strauss (A. K. Peters Ltd, 2008). Both texts aim at a level similar to this book, and Stewart's book introduces group theory in far greater depth than I did in Chapter 6.

A very readable, popular, and at the same time in-depth account of a geometrical approach to Einstein's general relativity is *General Relativity from A to B* by R. Geroch (University of Chicago Press, 1981). R. Penrose's *The Road to Reality* (Vintage Books, 2007) will take you all the way from vectors and calculus to Riemannian geometry, general relativity, and twistor theory.

Index

A

Agnesi, Maria 35
Al-Khujandi 118
angle 14
  opposite 11, 23
Apollonius of Perga 92
Archimedes of Syracuse 22, 51
area 31, 50, 72
  of a circle 34
  of a pseudosphere 64
  of a triangle 32, 60
Atiyah–Singer index theorem 103
atlas 75
axioms 1
  of a group 107
  of area 32, 33, 35
  of distance 16
  of Euclidean geometry 9
  of hyperbolic geometry 63
  of vectors 41

B

Beltrami Eugenio 54, 63
  disc model 65
Big Bang 139
black hole 133, 135
Bolyai Janos 11, 63
Brianchon Charles 99
  theorem of 5, 99

C

Cartan Elie 128
causal structure 123, 130
Ceva Giovanni 105
  theorem of 106
circle 45
  circumference 8
  great circle 47, 50, 52, 56
  on a surface 71
  osculating 68
compass and ruler 12, 20, 24, 37, 39, 92, 140
congruent 17, 32, 104
conic 92, 95, 98, 105
coordinates 138
  Cartesian 36, 74, 91, 114, 119, 121
  homogeneous 89, 100
Cosmic Censorship Hypothesis 140
cross-ratio 112
curvature 4, 65, 129
  of a curve 67
  extrinsic 71
  Gaussian 66, 70, 78, 102, 113
  intrinsic 71
  negative 71, 81
  of a surface 69
  positive 72, 82
  Ricci 131
  Riemann 81, 138
  Weyl 131, 132

D

da Vinci Leonardo 4, 42, 101
Desargues Girard 4, 88, 95, 101
  theorem of 88
Descartes Rene 36
dimension 38
  of a manifold 74
distance 15, 49, 57, 78, 102, 105, 128, 131
  hyperbolic 56

distance (*cont.*)
Minkowski 120
positivity 121
ruler 15, 17
duality
Platonic solids 31
projective 97, 141
Dürer Albrecht 85, 101

E

Einstein Albert 77, 121
equations 131, 132
Erlangen programme 104, 108
Escher M. C., 3, 58
Euclid of Alexandria 1, 9, 42
Euclid's fifth axiom, *see* parallel postulate
Euler Leonhard 30, 118
formula of 30
event 120, 123
event horizon 134

F

Faltings Gerd 118
Fermat's Last Theorem 114
Fields Medal 6, 114, 118

G

Gauss Carl Friedrich 4, 11, 24, 66, 103
Theorema Egregium 66, 71
Gauss–Bonnet theorem 73
Gaussian curvature *see* curvature
general relativity 6, 129, 131, 134, 139
genus 73
geodesic 80, 81, 102, 130, 133
incomplete 138
geometry 1, 103
affine 105
algebraic 5, 114, 118
analytic 37
differential 4, 68, 104
Euclidean 3, 9, 35, 72
hyperbolic 3, 54, 57, 60, 72
Minkowski 120
non-Euclidean 47, 63
projective 5, 83, 91, 109, 127
Riemannian 78
spherical 3, 48, 60, 72
synthetic 17
Greek 42
conics 92
cosmology 128
geometry 8
philosophy 29
problems of antiquity 20
group 107
Abelian 108
affine 109
Euclidean 108
of transformations 107
projective 110

H

Hawking Stephen 139

I

invariant 112
isometry 16, 50, 108, 126

K

Klein Felix 108, 111

L

Lambert Johann 60
light cone 122, 133
line 9, 50, 105
at infinity 86
hyperbolic 54, 65
in Minkowski space 123
number line 15

segment trisection 12
spherical 51
world line 124
Liouville Roger 102
Lobachevsky Nikolai 11, 63

M

manifold 52, 74, 103, 130, 132, 138, 140
patching conditions 76
Riemannian 80, 81
map 50, 51, 75
Mercator Gerardus 50
metric
British Rail 18
metrizability problem 102
pseudo-Riemannian 130, 131
Riemannian 79, 80, 113
Schwarzschild 136
Milnor John 77
Möbius August 111
Minkowski Hermann 121

N

Nash John 80
numbers 15, 19
chromatic 45
constructible 21
integers 20
irrational 20
transcendental 22

P

parallel postulate 10, 47, 54, 62
Pascal Blaise 95, 101
theorem of 95
Penrose Sir Roger 26, 138
aperiodic tiles 26
Penrose–Hawking singularity theorems 82, 139
Perelman Grigori 6, 113
perspective drawing 85
pi 7, 22
irrationality 22
Plato 20
Platonic solids 28
Poincaré Henri
conjecture 6, 113
disc model 54, 63
polygon 22
constructible 24
convex 23
hyperbolic 61
regular 24
polyhedron 28
Poncelet Jean-Victor 95, 101
theorem of 96
projective line 83, 111
projective plane 83, 90, 110
proof
Brianchon theorem 99
Desargues theorem 89
irrationality of $\sqrt{2}$, 19
Pythagoras theorem 1
Sylvester's problem 44
trisecting a line segment 13
pseudosphere 64
Pythagoras 1, 117
Pythagorean triple 115
theorem of 1, 38, 46

Q

quasi-crystals 28

R

Riemann Bernard 78

S

Schwarzschild Karl 134
Shechtman Dan 28
simultaneity 124
singularity 136, 138
space-time 121
Aristotelian 128

space-time (*cont.*)
  Galilean 128
  Newtonian 128
  of general relativity 129
  Schwarzschild 138
space-time interval 120
special relativity 121, 129
sphere 36, 38, 47
  celestial 127
  exotic 77
Steinhaus Hugo 119
stereographic projection
    75, 127
Sylvester James Joseph 43

T

tensor 81, 131
Thales of Miletus 8
tiling 25, 26, 45
  aperiodic 26
topology 112
transformation 105
  affine 18, 105, 107
  linear 109
  Lorentz 126
  Möbius 111, 127
  projective 110
triangle 22, 35
  hyperbolic 59
  ideal 60
  in perspective 89
  inequality 16, 18, 58
  spherical 50
  sum of internal angles 22, 48,
    53, 59
twistor theory 140

V

vector 40, 107
  tangent 78
volume 7

W

Wantzel Pierre 14
Wiles Sir Andrew 5, 114, 118

NUMBERS

A Very Short Introduction

Peter M. Higgins

Numbers are integral to our everyday lives and feature in everything we do. In this *Very Short Introduction* Peter M. Higgins, the renowned mathematics writer unravels the world of numbers; demonstrating its richness, and providing a comprehensive view of the idea of the number. Higgins paints a picture of the number world, considering how the modern number system matured over centuries. Explaining the various number types and showing how they behave, he introduces key concepts such as integers, fractions, real numbers, and imaginary numbers. By approaching the topic in a non-technical way and emphasising the basic principles and interactions of numbers with mathematics and science, Higgins also demonstrates the practical interactions and modern applications, such as encryption of confidential data on the internet.

www.oup.com/vsi

STATISTICS

A Very Short Introduction

David J. Hand

Modern statistics is very different from the dry and dusty discipline of the popular imagination. In its place is an exciting subject which uses deep theory and powerful software tools to shed light and enable understanding. And it sheds this light on all aspects of our lives, enabling astronomers to explore the origins of the universe, archaeologists to investigate ancient civilisations, governments to understand how to benefit and improve society, and businesses to learn how best to provide goods and services. Aimed at readers with no prior mathematical knowledge, this *Very Short Introduction* explores and explains how statistics work, and how we can decipher them.

www.oup.com/vsi

RELATIVITY
A Very Short Introduction
Russell Stannard

100 years ago, Einstein's theory of relativity shattered the world of physics. Our comforting Newtonian ideas of space and time were replaced by bizarre and counterintuitive conclusions: if you move at high speed, time slows down, space squashes up and you get heavier; travel fast enough and you could weigh as much as a jumbo jet, be squashed thinner than a CD without feeling a thing - and live for ever. And that was just the Special Theory. With the General Theory came even stranger ideas of curved space-time, and changed our understanding of gravity and the cosmos. This authoritative and entertaining *Very Short Introduction* makes the theory of relativity accessible and understandable. Using very little mathematics, Russell Stannard explains the important concepts of relativity, from E=mc2 to black holes, and explores the theory's impact on science and on our understanding of the universe.

www.oup.com/vsi

INFORMATION
A Very Short Introduction
Luciano Floridi

Luciano Floridi, a philosopher of information, cuts across many subjects, from a brief look at the mathematical roots of information - its definition and measurement in 'bits'- to its role in genetics (we are information), and its social meaning and value. He ends by considering the ethics of information, including issues of ownership, privacy, and accessibility; copyright and open source. For those unfamiliar with its precise meaning and wide applicability as a philosophical concept, 'information' may seem a bland or mundane topic. Those who have studied some science or philosophy or sociology will already be aware of its centrality and richness. But for all readers, whether from the humanities or sciences, Floridi gives a fascinating and inspirational introduction to this most fundamental of ideas.

'Splendidly pellucid.'

Steven Poole, The Guardian

www.oup.com/vsi

INNOVATION

A Very Short Introduction

Mark Dodgson & David Gann

This *Very Short Introduction* looks at what innovation is and why it affects us so profoundly. It examines how it occurs, who stimulates it, how it is pursued, and what its outcomes are, both positive and negative. Innovation is hugely challenging and failure is common, yet it is essential to our social and economic progress. Mark Dodgson and David Gann consider the extent to which our understanding of innovation developed over the past century and how it might be used to interpret the global economy we all face in the future.

'Innovation has always been fundamental to leadership, be it in the public or private arena. This insightful book teaches lessons from the successes of the past, and spotlights the challenges and the opportunities for innovation as we move from the industrial age to the knowledge economy.'

Sanford, Senior Vice President, IBM

www.oup.com/vsi

GALAXIES

A Very Short Introduction

John Gribbin

Galaxies are the building blocks of the Universe: standing like islands in space, each is made up of many hundreds of millions of stars in which the chemical elements are made, around which planets form, and where on at least one of those planets intelligent life has emerged. In this *Very Short Introduction*, renowned science writer John Gribbin describes the extraordinary things that astronomers are learning about galaxies, and explains how this can shed light on the origins and structure of the Universe.

www.oup.com/vsi